War and Strategy in the Modern World

This volume brings together some of Professor Azar Gat's most significant articles on the evolution of strategic doctrines and the transformation of war during the twentieth and early twenty-first centuries.

It sheds new light on the rise of the German Panzer arm and the doctrine of Blitzkrieg between the two world wars; explores the factors behind the formation of strategic policy and military doctrine in the world war era and during the Cold War; and explains why counterinsurgency has become such a problem. The book concludes with the spread of peace in the developed world, challenged as it is by the rise of the authoritarian-capitalist great powers – China and Russia – and by the chilling prospect of unconventional terrorism. This last chapter summarizes the author's latest research and has not previously been published in article form.

This collection will be of much interest to students of strategic studies, military history, and international relations.

Azar Gat is Ezer Weitzman Professor of National Security at Tel Aviv University. He is the author of eight books, including *A History of Military Thought: From the Enlightenment to the Cold War* (2001); *War in Human Civilization* (2006); *Victorious and Vulnerable: Why Democracy Won in the 20th Century and How It Is Still Imperiled* (2010); and *The Causes of War and the Spread of Peace: But Will War Rebound?* (2017). His books have been translated into Spanish, Japanese, Chinese, Korean, Greek, Turkish, and Hebrew.

Cass Military Studies

War and Strategy in the Modern World

From Blitzkrieg to Unconventional Terrorism

Azar Gat

Routledge
Taylor & Francis Group

LONDON AND NEW YORK

First published 2018
by Routledge

2 Park Square, Milton Park, Abingdon, Oxfordshire OX14 4RN
52 Vanderbilt Avenue, New York, NY 10017

Routledge is an imprint of the Taylor & Francis Group, an informa business

First issued in paperback 2020

British Library Cataloguing-in-Publication Data
A catalogue record for this book is available from the British Library

Library of Congress Cataloging-in-Publication Data
A catalog record for this book has been requested

ISBN: 978-1-138-63256-1 (hbk)
ISBN: 978-0-367-66698-9 (pbk)

Typeset in Times New Roman
by Apex CoVantage, LLC

Contents

Preface

This volume brings together a select collection of my articles on the evolution of strategic doctrines and the transformation of war during the twentieth and early twenty-first centuries. The articles are arranged in chronological order: the volume begins with the rise of the German Panzer arm and the doctrine of Blitzkrieg during the interwar period; it ends with the spread of peace in the developed world, threatened as it is by the challenge posed by the authoritarian-capitalist great powers – China and Russia – and by the chilling prospect of unconventional terrorism. The sequence of the articles also reflects the development of my interests and scholarly pursuits, starting with my books on the evolution of military theory and doctrine, assembled in the omnibus edition *A History of Military Thought: From the Enlightenment to the Cold War* (2000). My interests later expanded to broader questions relating to war, pursued in my *War in Human Civilization* (2006), *Victorious and Vulnerable: Why Democracy Won in the 20th Century and How It is still Imperiled* (2010), and *The Causes of War and the Spread of Peace: But Will War Rebound?* (2017).

In revising these articles for publication, I have tried not to make changes that would reflect the benefit of hindsight. Indeed, I have been satisfied to see that the articles have well withstood the test of time. Choosing to leave them basically in their original form has also meant that later publications on their respective subjects have not been included either. I do not find that they significantly alter my conclusions. Most of the changes I have introduced are cuts made to avoid repetitions and overlaps between the articles. I have also made occasional stylistic changes.

Misconceptions regarding the rise of the Panzer arm which gave Germany its lightning victories at the beginning of World War II are, surprisingly, very significant indeed. 'British Influence, the Evolution of the Panzer Arm, and the Rise of Blitzkrieg' sets out to correct some of these major misconceptions. Revelations that the famous British military theorist B.H. Liddell Hart manipulated the German generals' testimonies after World War II have discredited the claim of a decisive British influence on the evolution of the German Panzer arm during the 1920s and 1930s. However, the German archives from that period reveal that this influence was indeed paramount. The article also shows that fateful historical accidents, unnoticed by scholars, were largely responsible for the fact that the Panzer arm did not follow the mistaken routes taken by the other great powers with respect to the organization of armour. Finally, the article shows that, rather than being a formal doctrine formulated by the German armed forces during the 1930s, 'Blitzkrieg' emerged as an operational concept only during the early campaigns of World War II, while the word itself was sensationally coined by the foreign press.

In the wake of World War II, the controversies of the interwar period regarding both national policy and strategic doctrine were dramatically viewed as struggles between prescience and folly. This narrative continues to dominate the popular view and the media. However, from the late 1960s, as national archives opened, scholars have been formulating a more complex and nuanced picture, in which 'right' and 'wrong' have not been as starkly contrasted as before. In 'Technology, National Policy, Ideology, and Strategic Doctrine between the World Wars', I outline the real dilemmas, deep constraints, genuine uncertainties, and conflicting goals which haunted governments and military establishments during the 1930s.

'Isolationism, Appeasement, Containment, Limited War: The Democracies' Strategic Policy from the Modern to the "Post-Modern" Era' was my first article on the question of whether or not modern liberal democracies were special and different from other societies and regimes in their conflict behaviour. The article argues that, ever since the beginning of the twentieth century, democratic great powers have tended to follow a characteristic strategic pattern in the face of threats. They move cautiously up the scale from appeasement, to containment and cold war, to limited war, and only most reluctantly to full-fledged war. This sequence underlays the democracies' response to the German, Japanese, and Soviet challenges alike before and during the three great power clashes of the twentieth century. And it is still evident in the democracies' policies towards both strong and weak rivals in today's world.

The conspicuous changes that have taken place in the character of warfare over the past decades have been titled the 'Revolution in Military Affairs' (RMA). The problem with this label, however, is that it tells us nothing about the nature of the revolution and its place in the broader sweep of technology-driven revolutions of the industrial-technological age. The article 'The "Revolution in Military Affairs" (RMA) Compared with Earlier Military-Technological Revolutions of the Nineteenth and Twentieth Centuries' addresses this broader context. Over the past two centuries, innovations in technology accelerated dramatically in comparison to pre-industrial times, with military technology constituting only one aspect of this general trend. In close unison with civilian developments, military technology has undergone three major revolutionary waves, corresponding to and closely matching the characteristics of the first, second, and third (electronic-information) industrial-technological revolutions.

From earliest times and throughout history, fighting has been associated with men. Cross-cultural studies of male-female differences have found that serious violence is the most distinctive sex-related behavioural difference. Is this difference a matter of education and social conventions, or are men naturally far more adapted to fighting than women are? This question is at the centre of a heated public debate regarding women's equality in modern society: can and should women enlist in combat roles in the armed services? 'Female Participation in War: Bio-Cultural Interactions' attempts to elucidate the respective roles of nature and nurture in this question, whose complexity, and even existence, are all too often ignored in this debate. This may facilitate a realistic, cool-headed and non-ideological assessment of the possibilities and of future trends.

The last five chapters in this volume return, from various angles, to the question of how and why modern liberal democracies differ in their conflict and war behaviour from other societies and regimes. For example, the liberal democracies' colonial record includes a particularly problematic element: the charge that in both the United States and Australia, democracies exterminated the native populations. This is the source of a profound sense of guilt in the two countries, reinforcing pervasive doubts about whether liberal democratic societies really behave better than others. In his book *The Dark Side of Democracy: Explaining Ethnic Cleansing* (2004), historical sociologist Michael Mann suggests that the democracies are particularly prone to genocide. However, in 'Is Democracy Genocidal?' I show that this charge is fundamentally invalid, as are the conclusions drawn from it.

Down to the wars in Afghanistan and Iraq, insurgency warfare has earned a reputation of near invincibility, driving great powers out of their former colonial empires during the twentieth century and frustrating military interventions even where the asymmetry in regular force capability is the starkest. Why have mighty powers that proved capable of crushing the strongest of opponents failed to defeat the humblest of military rivals in some of the world's poorest and weakest regions? Composed in collaboration with Gil Merom, 'Why Counterinsurgency Fails' argues that, rather than being universal, this difficulty has overwhelmingly been the lot of liberal democratic powers – and encountered precisely because they are liberal and democratic. The crushing of an insurgency necessitates ruthless pressure on the civilian population, which modern liberal democracies have found increasingly unacceptable. Premodern powers, as well as modern authoritarian and totalitarian states, have rarely had problems with such measures, and overall they have proved quite successful in suppression. The measures proposed in this article for fighting guerrilla given liberal societies' norms and sensibilities – a reliance on local allies on the ground and extensive use of high-tech, stand-off, accurate fire, aircraft (manned and unmanned), and special forces to minimize friction with the civilian population – have since become the methods of choice for the democracies' conduct in such operations. At the same time, the article highlights the inherent limitations of these measures.

Democracy emerged victorious from all the great power struggles of the twentieth century – the two world wars and the Cold War – surviving both its right-wing and left-wing authoritarian and totalitarian rivals. To many, most famously Francis Fukuyama, this suggested some inherent selective advantages for democracy, conferring an air of inevitability on the past as well as on the future. 'The Return of the Authoritarian-Capitalist Great Powers: Is the Democratic Victory Preordained?' addresses the question of why the democracies won in the past, and what this can teach us about the future. The article argues that whereas the communist great powers, the Soviet Union and China, lost because they indeed proved to be economically inefficient, the capitalist nondemocratic great powers, Germany and Japan, were defeated because they happened to be too small to contend with continental-size giants, most notably the United States. This analysis is relevant to the twenty-first century. Today's China (and to a lesser degree a territorially and demographically much reduced Russia) is the giant in the system, which for long was held back by its inefficient communist economy. However, over the past decades it has transitioned to a much more efficient, and hence more powerful, form of authoritarianism. We thus face a new, historically unprecedented challenge – a nondemocratic superpower which is both big and capitalist. The main part of this article was written and published before the outbreak of the Great Recession in 2007–2008, when the euphoria and triumphalism surrounding capitalist liberal democracy were still pervasive. Since then, the lustre of liberal democracy has dimmed, the challenge from both China and Russia has become more overt, and the Third Wave of democratization has stalled. As the article argues, while it may well be that China and Russia would eventually liberalize and democratize, this should be regarded as an open question rather than a necessary outcome of socio-economic development, as the prevailing reading of twentieth century history held at the time.

The Arab Upheaval has been the cause of profound bewilderment in the West and among policy makers. Great enthusiasm for the Arab Spring was quickly replaced by confusion and concern regarding an Islamist Winter. And this was as quickly supplanted by disconcert and despair in the face of military takeovers and ferocious civil wars. The European revolutions of 1848, the Spring of Nations, with their great hopes and dashed dreams, have often been cited as an analogue. 'A Compass to the Arab Upheaval: What Can Nineteenth Century Europe Teach?' asks what the European experience of modernization and regime change during the nineteenth and early twentieth centuries may suggest with respect to the contemporary Arab world. While history does not quite repeat itself, it is still the best guide we have. The article cautions against unrealistic expectations and a historically insensitive application of ideological abstractions to the Arab world, its level of development and existing social and cultural characteristics. It was written before the rise of the Islamic State (ISIS) which has since captured the headlines. But its conclusions remain unchanged.

International relations theorists have identified a number of peace phenomena, most notably the democratic/liberal peace and commercial/capitalist peace. However, the historical record reveals gaps and inconsistencies with the Kantian formula for peace: premodern democracies and republics did fight each other; until the nineteenth century, rather than trade peacefully, states tried to monopolize trade by force; nondemocratic countries, and not only democracies, have participated in the general decrease in belligerency during the past two centuries, including communist powers that largely opted out of the global trade system. 'The Modernization Peace and Twenty-first Century Conflict' sets out to explain these problems in the prevailing peace theories, and at the same time reconcile, unify and transcend these theories into a broader whole. It compresses into article form my writings on the subject over the past decade and in my most recent book. The article argues that the process of modernization – unfolding since the onset of the industrial age in the early nineteenth century and constituting the greatest revolution in human history, but practically ignored in international relations theory – is the substratum on which the various peace phenomena ride. Hence the marked decline in belligerency since 1815 among both democratic and nondemocratic, and capitalist and non-capitalist, countries (albeit at different rates). Rather than war becoming more lethal and expensive under modern conditions (it hasn't), it is actually peace that has become more rewarding. Finally, the article explains the great divergence from the trend, the world wars, and explores how the various elements of the Modernization Peace might unfold in the twenty-first century and how this peace may still be challenged. Threats include alternative modernizers, such as today's China and Russia, and anti-modernists and failed modernizers that may spawn terrorism, potentially unconventional. The world has become more peaceful than ever before, with both inter- and intra-state war disappearing from its most developed and affluent parts, the areas most affected by the Modernization Peace. And yet there is still much to worry about in terms of security and there is no place for complacency.

Acknowledgements

This volume consists of previously published essays and the author would like to acknowledge the permissions granted to reprint these articles and book chapters.

1 Azar Gat, 'British Influence and the Evolution of the Panzer Arm – Myth or Reality?', *War in History*, 4, 1997, 150–173, 316–338. Reprinted by permission of SAGE Publishing.
2 Azar Gat, 'Ideology, National Policy, Technology and Strategic Doctrine between the World Wars', *The Journal of Strategic Studies*, 24 (3), 2001, 1–18. Reprinted by permission of Taylor & Francis.
3 Azar Gat, 'Isolationism, Appeasement, Containment, Limited War: Western Strategic Policy from the Modern to the "Post-Modern" Era', in Zeev Maoz and Azar Gat (eds.), *War in a Changing World*, Ann Arbor: University of Michigan Press, 2001, 77–91. Reprinted by permission of the University of Michigan Press.
4 Azar Gat, 'The "Revolution in Military Affairs" (RMA) as an Analytical Tool for the Interpretation of Military History', in D. Adamsky and K. Bjerga (eds.), *Contemporary Military Innovation*, New York: Routledge, 2012, 7–19. Reprinted by permission of Taylor & Francis.
5 Azar Gat, 'Female Participation in War: Bio-Cultural Interactions', *The Journal of Strategic Studies*, 23 (4), 2000, 21–31. Reprinted by permission of Taylor & Francis.
6 Azar Gat, 'Is Democracy Genocidal?', review article of Michael Mann, 'The Dark Side of Democracy', *The Times Literary Supplement*, 21 March 2007. Reprinted by permission of the publishers.
7 Azar Gat and Gil Merom, 'Why Counterinsurgency Fails?', in Azar Gat, *Victorious and Vulnerable: Why Democracy Won in the 20th Century and How It Is Still Imperiled*, Hoover Institution, Stanford/ Rowman and Littlefield, 2010. Reprinted by permission of Rowman and Littlefield.
8 Azar Gat, 'The Return of Authoritarian Great Powers', *Foreign Affairs*, 86 (4), 2007, 59–69; idem., 'Are Authoritarian China and Russia Doomed? Is Liberal Democracy's Victory Preordained?', *Foreign Affairs*, 88 (3), 2009. Reprinted by permission of the publishers.
9 Azar Gat, 'The Arabs' 1848', *The National Interest*, 20 April 2014. Reprinted by permission of the publishers.

1 British influence, the evolution of the Panzer arm, and the rise of Blitzkrieg

The British influence – a fraud?

B.H. Liddell Hart's reputation as one who decisively influenced the proponents of armoured warfare in Germany during the interwar period has been marred and thrown into question by revelations that this reputation was largely self-propagated, and that to create it he actually exploited the plight in which the German generals were after the Second World War, unscrupulously manipulating their evidence for his own ends. His personal contacts with the German generals, his role as the one who recorded and presented their war histories to western readers, and his strong public support for them bound the generals to him by feelings of gratitude, self-interest, and dependency.[1] John Mearsheimer has fully exposed Liddell Hart's persistent efforts and elaborate techniques in using his connections with the German generals for extracting, inviting, and planting accolades, which he later inflated beyond their original context, modified, inserted in key publications, and disseminated widely by any possible means. As Mearsheimer has shown, three cases were of particular significance for Liddell Hart: Hans Guderian, Erwin Rommel, and Erich Manstein.

In Manstein's case Liddell Hart's efforts did not bear fruit, despite the fact that the field marshal was heavily in his debt. Liddell Hart intervened to relieve the hardship and humiliation which Manstein endured in a prisoner-of-war camp. On Manstein's request he arranged for his wife and child to be transferred to his sister's house in the French zone of occupation in Germany. He campaigned against Manstein's being tried as a war criminal, assisted in his defence when he was put on trial, and fought for his release after he had been convicted. He took under his care the publication of the English edition of Manstein's war memoirs, *Lost Victories* (1958), and as late as the 1960s intervened to secure a place at Cambridge for Manstein's son. Yet, despite Manstein's gratitude, he withstood Liddell Hart's attempts to make the latter the inspiration behind the Ardennes operation which Manstein had conceived and which had led to the Allies' collapse in the West in 1940. This, however, did not stop Liddell Hart from putting his words in Manstein's mouth in his *Memoirs*.[2]

Liddell Hart had more luck with Rommel's family. The field marshal's widow and son were very anxious that Liddell Hart would prepare an English edition of his papers. On his persistent urging, Rommel's family and his chief of staff in North Africa, General Fritz Bayerlein, provided flimsy but reasonable evidence that Rommel, like most German officers, had known of Liddell Hart during the 1930s and had probably read some of his writings, though Rommel himself had not been converted to armour before 1940. The evidence further showed that during the war Rommel had on two different occasions mentioned the failure of his British opponents to adopt the theories of armoured warfare originally developed by 'British military critics' (Bayerlein explained that Rommel had meant [J.F.C.] Fuller and Liddell Hart). In one of his papers Rommel had also specifically referred to an article Liddell Hart had written during the war. Liddell Hart, however, only accepted the job of editing Rommel's papers after extracting from Rommel's family and from Bayerlein statements that made Rommel nothing less than his 'pupil' who had been 'highly influenced by his tactical and strategic conceptions'. He inserted this statement in the English edition of *The Rommel Papers* (1953), but failed to make Bayerlein have it incorporated in the German one.[3]

The most important case for Liddell Hart, and the one in which he achieved his crowning success, was that of Guderian, Germany's foremost armour pioneer. The two corresponded extensively from September 1948. The brisk and abrasive Guderian had made himself quite a number of enemies in the German army, and was interesting in getting his side of the story told. Liddell Hart's interviews with the German generals, *The Other Side of the Hill* (1948), had been published before he and Liddell Hart made contact, but Liddell Hart was planning a second, enlarged edition of the book. Six months after they began their correspondence he informed Guderian that he intended to devote a whole chapter to him in the new edition. At about the same time he inquired if Guderian had considered writing his war memoirs. Guderian, who was receiving no pension, was then living with his wife in one room under conditions of virtual poverty. As he wrote to Liddell Hart, publishing his memoirs was, if nothing else, a means for him to earn a living.[4] Liddell Hart took it upon himself to find British and American publishers for the memoirs and also put Guderian in touch with British and American journals. Getting the memoirs accepted for publication in the West proved, however, very difficult. Two publishing houses, Collins and Cassell, successively rejected the typescript, describing it (rightly) as 'full of self-pity and unrepentant nationalism, typical of a German officer of the nationalistic school'.[5] Liddell Hart worked hard to soften and remove the problematic passages in the book, find another publisher, and, finally, secure the best financial terms for Guderian. When the book, *Panzer Leader* (1952), became a bestseller, he asked for the 25 per cent of the royalties which Guderian himself had offered him for his immense trouble. His request remained unanswered, for Guderian had just died.

As Mearsheimer has pointed out, the more Guderian's debt to Liddell Hart had grown, the more persistent Liddell Hart's enquiries became regarding his influence upon Guderian, and the more Guderian realized that he would have to contribute the kind of acknowledgement that Liddell Hart wanted to maintain the mutually beneficial relationship. When Guderian failed to respond to hints, Liddell Hart resorted to more direct measures. In the German edition of his memoirs, *Erinnerungen eines Soldaten* (1951), Guderian wrote the following paragraph:

> It was principally the books and articles of the Englishmen, Fuller, Liddell Hart and Martel, that excited my interest and gave me food for thought. These far-sighted soldiers were even then trying to make the tank more than just an infantry support weapon. They envisaged it in relation to the growing mechanization of our age, and thus they became the pioneers of a new type of warfare on the largest scale.

Going over the English translation of the book, Liddell Hart was unsatisfied. He wrote to Guderian:

> I appreciate very much what you said in the paragraph. . . . So I am sure will Fuller and Martel. It is a most generous acknowledgement. But because of our special association and the wish that I should write the foreword to your book, people may wonder why there is no separate reference to what my writings taught. You might care to insert a remark that I emphasized the use of armoured forces for long-range operations against the opposing army's communications, and also proposed a type of armoured division combining panzer and panzer-infantry units – and that these points particularly impressed you.

Coming after Liddell Hart's tremendous efforts over the publication and contract of the book, this request was not refused. Guderian inserted the substantive sentences of Liddell Hart's letter in *Panzer Leader* after the paragraph he had originally written for the German edition.[6]

Guderian's lavish acknowledgement established Liddell Hart's reputation for a generation as the inspiration behind the German Blitzkrieg. Such strong evidence left little room for doubt, especially as Liddell Hart took care to cover his tracks. He apparently removed his letter to Guderian and Guderian's letter of agreement from his archive. Only in the mid-1970s were the incriminating letters discovered in Guderian's records by his biographer, Kenneth Macksey, and replaced back in the archive.[7]

When manufactured evidence is revealed, the damage to one's case might be fatal. Liddell Hart's claim for influence on the Germans has lost credibility in the eyes of historians. At the very least it has become clear that he exaggerated this influence at the expense of Fuller and other British armour pioneers. At the same time, not only his significance but British influence as a whole on the evolution of the German Panzer arm, which was previously taken for granted, has now been called into question. And yet Liddell Hart's self-inflicted injury does not close the case, but merely opens it afresh. The fact that he was fraudulent does not necessarily mean that he was wholly incorrect. To establish how things really were, the evidence on the subject from the German sources of the interwar period itself must be looked into. This has simply never been done. Liddell Hart himself did not read German, and he was anyhow satisfied with what he had managed to extract from the German generals directly. His biographers too confined themselves solely to his own records. Only in recent years have historians, working on the other side of the hill on other subjects using the German documents, dug up some evidence relevant to our case. Although a great deal of the German archival material was destroyed by the war or lost, leaving considerable gaps in the record, the surviving material is substantial. In addition, open publications from the interwar period, particularly the general staff's semi-official *Militär-Wochenblatt*, a professional journal of high quality, provide a very useful and often parallel source which complements the official record.

There are many parallels between the current trends in the historiography of interwar British and German armour. As with Fuller and Liddell Hart, it has become apparent that Guderian monopolized the history of the Panzer arm. The many existing popular histories of the development of German armour merely paraphrase Guderian's *Panzer Leader*, and his biographers have not diverged from his own version either.[8] Surprisingly for a subject that has attracted so much interest, a full-scale scholarly history of the German Panzer arm, based on the documents, has yet to be written. This study, of course, can fill the gap only partly. It will attempt to outline the genesis of the Panzer arm and the growth of its operational doctrine, with special attention to the British influence on these developments, including that of Liddell Hart. As will be shown, this influence was indeed, after all, decisive.

British pioneers and the Reichswehr's awakening interest in mechanized warfare

The first substantial modification to have been made in Guderian's version concerns the notion that the German army's serious interest in armour was born, and had always been associated, with him. This is very far from the truth. Contrary to its popular image as professionally conservative, an image fostered by Guderian's memoirs, the Reichswehr showed lively interest in armour in the 1920s. At that time Guderian was only beginning to develop as an armour man and was still remote from positions of influence. A study of the Reichswehr in the Seeckt era has recently highlighted all this in considerable detail.[9] Compared with the Entente powers, Germany created a very small tank force only late in the First World War, and came out of the war with little practical experience in tank warfare. The stipulations of the Versailles Treaty, which prevented Germany from building and possessing tanks, further fundamentally hindered her development in the field of armour. For both reasons, however, the Reichswehr's sensitivity to the new weapon was in some respects heightened. For both reasons it was also especially conscious of, and dependent upon, developments abroad to a degree that no other great power's army was.

Thus the German postwar field service regulations, *Leadership and Battle with Combined Arms* (1923), dedicated several sections to the tank. Taking their cue from the postwar doctrines of the French and British armies, the regulations incorporated advanced ideas regarding the use of heavy and light tanks for the break-in and cavalry-type missions.[10] The sections on the tank in the regulations were probably drafted by Lieutenant Ernst Volckheim, a veteran of the First World War German tank force, who served after the war in the inspectorate of transport troops (In.6). They are practically identical to his own publications. Volckheim was well recognized towards the middle of the 1920s as the Reichswehr's leading expert on the use of armour. He was well informed about the history of the tank in the First World War and the French and British postwar armour organization and doctrine, upon which he relied and which he introduced to German readers.[11]

Germany's foremost expert on tank technology and another major source of information on the world's armour for German readers was the Austrian captain, engineer Fritz Heigl. His *Pocket Book of Tanks* (1926) was the standard work on the subject which, expanded and updated by others after its author's premature death, ran into three editions before the war. Heigl also advised the German army on tanks.[12] In 1925–1926 the Reichswehr issued preliminary specifications for the building of two experimental tank types, to be produced by German firms in secrecy in order to avoid detection by the Allies. These were codenamed 'Heavy Tractor' and 'Light Tractor' – fast-heavy and medium-light types respectively. From 1929–1930 the various models produced were secretly tested in the joint German–Soviet tank school in Kama, near Kazan.[13] Both types possessed high speed (30–35 km/h) and resembled the British Independent and Vickers Medium Tank respectively.

Indeed, at the time these models were launched attention in Germany was increasingly focusing on new and exciting developments in armoured warfare coming from Britain. By 1924–1925, in his last publications before disappearing from the scene, Volckheim, who had previously been more influenced by the larger and closer French tank force, was beginning to whistle new tunes. He cited a British officer's criticism of the French army's lack of a special inspectorate for tanks, and a British claim, following the latest French manoeuvres, that the French had made no progress in tank warfare since the war. He called attention to the heavily armed and fast (British) medium tank as a new significant development in the field, and described its use in the 1924 British manoeuvres in cooperation with armoured cars and cavalry. In 1926 Heigl's survey of the world's tanks mentioned the new theories, specifically associated with Fuller, of using fast tanks with a large radius of action to revive the war of movement.[14]

The earliest significant introduction of the new British school to German readers was Liddell Hart's articles 'The Next Great War' and 'The Development of a "New Model" Army: Suggestions on a Progressive, but Gradual Mechanization', published in 1924 in the *Royal Engineers Journal* and *The Army Quarterly* respectively. The former was abstracted as the opening piece of a July issue of *Militär-Wochenblatt*, whereas the latter received only a few lines in the regular military journals section in November but was described as 'very interesting for all concerned with the mechanization of modern armies' and recommended for a full translation into German.[15] Here, as always, the summaries in the German journal were accurate and to the point. A month later, without mentioning either Liddell Hart's name or Fuller's (from whom the former derived his ideas[16]), another opening piece in *Militär-Wochenblatt* described the new thoughts in Britain of replacing the muscle armies by machine armies through gradual mechanization in several phases, ending with an all-armoured army and a reduction of 60 per cent in manpower. The article concluded that this programme would be tested in the next British summer manoeuvres in 1925. As was often the case in *Militär-Wochenblatt*, the summary was contributed by a general staff officer whose field of speciality covered the subject reviewed, and who for reasons of confidentiality signed only with a number.[17]

So Liddell Hart was basically correct in claiming that his earliest articles on armour had left an impression in Germany.[18] Yet this impression should be understood within a wider context. The older historiography, taking its cue from the writings of Fuller, Liddell Hart, and Guderian, emphasized the role of individuals and theories in the evolution of armour. For all their significance, however, armies are even more impressed and spurred into action by tangible developments in reality. It was only in combination with the path-breaking developments in tank design and armour organization in Britain, which were actually being tested in large-scale manoeuvres, that the British pioneering theories of armoured warfare, which had been a necessary condition for these developments in the first place, attracted so much attention in Germany and in other great powers' armies.

In late 1924 the intelligence branch of the German covert general staff (Truppenamt) surveyed the previous summer's British manoeuvres. In the section dealing with tanks and motorized troops the survey highlighted the appearance of the new Vickers Medium Tank, armed with a three-pounder and capable of a revolutionary 35 km/h. The survey emphasized the tank's potential for use in a war of movement, and noted a tension in this respect between the older and younger officers in the British army, the latter regarding the new tank as 'almighty'. The motorization on lorries of some of the other forces was also noted.[19] *Militär-Wochenblatt* published a summary of the survey, making the same points. It was written by the same general staff officer who three weeks earlier had described in the journal the new thoughts in Britain regarding the employment of armour.[20]

In the following years close attention to British theory and practice was strongly evident in the German army. Reports on and references to the revolutionary characteristics of the Vickers Medium Tank were unceasing, intertwined with reports on the British manoeuvres.[21] In 'A Reflection on the Employment of the Tank', a German officer wrote that the development of the tank since the war, embodied in the medium tank's speed and radius of action, opened new possibilities beyond its use for infantry support. This, stated the article, was the opinion of the British Colonel Fuller. Future war would be conducted by mechanized divisions in which all arms would be mechanized and armoured.[22] Several months later the same officer reported an article on the armour manoeuvres in the *Daily Telegraph* by 'Captain Liddell Hart, known from his book *Paris*'. The report highlighted the potential of the modern tank as against the views of the 'old school'.[23] Yet another article, 'The Impact of Modern Tanks on the Conduct of War', compared the British tanks and the British manoeuvres to the French. It stressed the great value of fast mechanized formations, and described the British Medium Tank as the most advanced in the world, possessing amazing speed and radius of action. 'It has developed into the pure offensive weapon of the war of movement'. Wide-range drives have now become possible. The Germans, the article concluded, must take up the subject without delay.[24]

The acute awareness in Germany of the inextricable link between theories and practice in the development of British armour is apparent in the passages cited. A summary of Liddell Hart's *Paris* opened an issue of *Militär-Wochenblatt* in September 1926. His reports, particularly in the *Daily Telegraph*, were regularly cited.[25] An article by Tim Pile in the *Journal of the Royal Artillery* was summarized as 'British Views on the Development and Employment of Tanks', arguing that the tank would revolutionize war and stressing the differences from the French approach.[26] Fuller was slower to leave his mark in print in Germany, though experts were well aware both of his background with the Tank Corps in the First World War and of his paramount role in the new British school of armoured warfare. He was occasionally cited, but his earlier books were introduced to German readers somewhat belatedly. It was only his nomination in 1926 as assistant to the Chief of Imperial General Staff (CIGS), George Milne, with a special mandate for the advancement of mechanization that finally put him in the spotlight.[27] His *The Reformation of War* (1923) was now picked up and summarized in *Militär-Wochenblatt* and, extensively, in three issues devoted almost entirely to it, in the general staff's foreign military literature journal.[28] His *Tanks in the Great War* was translated sometime during the 1920s and circulated in typescript for internal use.[29]

To be sure, the Reichswehr was monitoring all armies and all foreign military publications on a grand scale. British armour developments and British manoeuvres were being reported, but so also were those of the French, Polish, Czechoslovakians, Soviets, Italians, Americans, Japanese, and at least a dozen smaller countries, inside and outside Europe. Articles and books by Fuller and Liddell Hart were summarized, but so were works by numerous other military writers from many countries.[30] And yet there was a difference. By 1925–1927 enthusiasm in the Reichswehr for the British advances in armoured warfare was reaching fever pitch, both in official and more popular circles. Fuller and Liddell Hart in particular were becoming household names.

In a confidential memorandum in May 1926 Werner von Blomberg, head of the operations branch in the German general staff, cited the British secretary of state for war's statement in Parliament that exercises had proved that the horse and the tank did not combine well; that the horse would be retained mainly for non-European theatres; that the British army would be gradually mechanized; and that a small mechanized force would be created immediately. The conclusion was: 'the British are the most advanced in mechanization.' A summary of the official document also appeared in *Militär-Wochenblatt*.[31] In October 1926 the British July manoeuvres, in which a regular infantry brigade was confronted by one composed of cavalry, mobile artillery, and a company of the Vickers Medium Tank, was reviewed in the journal. The review was based on Liddell Hart's reports in the *Daily Telegraph*. It also cited his own lessons from the manoeuvre, among them the maxim: 'move in dispersed order' (quoted in English); the importance of radio communication; and the conclusion that only mobile artillery can keep pace with the tanks.[32]

In November 1926 a memorandum signed by General Wilhelm Heye, Seeckt's successor as commander-in-chief of the German army (*Chef der Heeresleitung*), again elaborated on the potential of the new tanks as revealed in the manoeuvres (speed: 30–45 km/h; range: 193 km). The memorandum asserted that the French possessed a large number of tanks which were, however, obsolescent, dating back to the war, whereas the British had got rid of their old models. The British used their tanks to attack the enemy's rear, reserves, command posts, and artillery.[33] A change in German doctrine soon followed. 'By late 1926 a directive set forth that tank units could be separated from a "slowly moving infantry" and that tanks could be best used either in conjunction with "mobile troops or as independent units".' In January 1927 a memorandum signed by the head of the operations branch (T1) in the general staff, Werner von Fritsch, stated: 'Armoured, quickly moving tanks most probably will become the operationally decisive offensive weapon. From an operational perspective this weapon will be most effective if concentrated in independent units like tank brigades.'[34]

The climax of the new British advances in the field of armoured warfare and of the German interest in them was, of course, the first trials of the Experimental Mechanized Brigade in the summer of 1927. In April a report in *Militär-Wochenblatt* cited Liddell Hart's prescriptions for the brigade in the *Daily Telegraph*. The brigade, he wrote, ought to operate independently against distant targets. It should be used as the mobile reserve and decisive arm of manoeuvre in the hands of the commander-in-chief.[35] The manoeuvre itself was described in the journal in four consecutive issues, in full detail, day by day, complete with maps.[36] Milne's Tidworth address at the end of the manoeuvres was summarized in the general staffs foreign literature journal from Liddell Hart's report in the *Daily Telegraph*, and also cited in *Militär-Wochenblatt* from Liddell Hart's report in the *Journal of the Royal Tanks Corps*. Armoured divisions would be created in the future, promised the British CIGS. 'With the opening of a campaign the armoured force will penetrate 300–450 km deep into enemy territory' or circle around its strategic flank. The writer in *Militär-Wochenblatt* called attention to the address as expressing the views of the British general staff.[37]

The German journal also announced the publication in Britain and summarized the content of *Provisional Instructions for Tank and Armoured Car Training* (1927). As Guderian would write in reference to 1928, 'the current English handbook on armoured fighting vehicles was translated into German and for many years served as the theoretical manual for our developing ideas.' The manual was sold freely in Britain by His Majesty's Stationery Office. In due course, the Germans would also get hold of the confidential pioneering *Mechanized and Armoured Formations* (1929) and its successor, *Modern Formations* (1931).[38] Finally, reporting on the plans for the second season of the Armoured Force's trials in 1928, *Militär-Wochenblatt* cited a British source who predicted that the mechanization reforms would be carried through after the trials, which would signify 'victory for the mechanization school, for which writers like Colonel Fuller, Captain Liddell Hart and Colonel Rowan-Robinson have done so much'.[39]

In 1927 Liddell Hart's article in the *Daily Telegraph*, 'The Remaking of Modern Armies', was summarized in the general staff's foreign military literature journal. The book bearing the same title was extensively summarized a year later in two issues dedicated almost entirely to it, and also, in considerable detail, in *Militär-Wochenblatt*.[40] Fuller's *On Future Warfare* (1928) was summarized, a chapter each, in eight issues of the foreign literature journal, after an introductory review by Liddell Hart in the *Daily Telegraph* following its publication.[41] Articles by and references to the two British military writers regularly appeared in German journals.[42] Their books published in Britain were prominently reviewed in *Militär-Wochenblatt* (which was uncommon for books in foreign languages), often in special review articles rather than in the regular book review section, and normally in very complimentary terms.[43] Some criticisms, mainly of the ideas of Fuller, recognized as the source and senior figure of the new school, were also printed, though the general picture remained one of absorbing interest.[44] Articles by other British armour pioneers such as Ernest Swinton, Giffard le Q. Martel, Percy Hobart, Charles Broad, and Henry Rowan-Robinson, were also occasionally cited during the second half of the 1920s.[45] But Fuller and Liddell Hart were in a league of their own internationally, in both quantity of coverage and acclaim. An index of the general staff's foreign military literature journal shows Fuller and Liddell Hart occupying the lion's share of the section 'Future Warfare' in the late 1920s.[46]

At first, introductory references in German journals spoke of 'Captain Liddell Hart, known from his book *Paris*', or 'Captain Liddell Hart, who has made a name for himself in Britain as a military writer'; but soon it was to be always the formula 'the well-known Captain Liddell Hart', or simply 'Liddell Hart' – an old acquaintance who needed no introduction. A survey in *Militär-Wochenblatt* of an issue of the RUSI Journal described his article '1927 or 527' with the words: 'a brief presentation of the author's well-known views'.[47] Nothing more was needed. Things were similar with Fuller. General Günther Blumentritt's recollection of the period, though written after the Second World War, seems to capture the feeling of the time:

> Liddell Hart and Fuller were for us young officers after 1920 'the modern military authors'. Particularly in the Reichswehr they were carefully studied and all their articles read. In those days we were lieutenants and captains aged 28 to 35 and took delight in the modern spirit of these writers.[48]

Even to a greater degree than in Britain, the rise of Fuller and – more dramatically, considering his starting position – Liddell Hart from anonymity to fame in professional circles in Germany was meteoric, taking barely more than a couple of years.

All this was part of a wider trend. In the second half of the 1920s the theme of future warfare was second in prominence only to the First World War on the pages of *Militär-Wochenblatt*. The Reichswehr was entering new fields.

Following the German manoeuvres of 1928, the decision to convert the motor-ized transportation units into combat units was finally taken.[49] In 1929, as the chief of staff of the inspectorate of motorized troops, Colonel Oswald Lutz, was summarizing the data on mechanization in Britain,[50] Blomberg, now the chief of the general staff (Truppenamt), was drafting plans for the creation of mechanized formations.[51]

Indeed, by 1929, prompted by the British example which had been attentively watched by the general staffs of all the great powers, seminal trials in mechani-zation had been initiated by other armies as well. In 1928, on the instruction of the secretary of state for war, Dwight F. Davis, who had witnessed the British 1927 manoeuvres, the US Army formed and tried out for the first time its own experimental armoured brigade. The French were quickening the mechanization of their cavalry.[52] The Italians too were closely studying and much influenced by the British manoeuvres.[53] The first Soviet mechanized regiments were created in 1929, and the concept of 'Deep Battle' was evolved from 1928–1929 onward, stimulated initially by the British pioneering advances, though soon taking its own interesting and largely independent course. Blomberg, who visited the Red Army in 1928, was deeply impressed by the vigour of the Soviet state, though, significantly for our case, criticized the Soviet tanks for lack of sufficient speed.[54] It was to the leading British model that German eyes were turned.

In view of Blomberg's prominent involvement in the developments described, it comes as no surprise that at the Disarmament Conference in Geneva, where he headed the German military delegation, he asked to meet Liddell Hart and told him of his admiration for his work. For obvious reasons Blomberg did not mention the British influence on the clandestine evolution of German armour. Instead, he discussed with Liddell Hart the latter's ideas in *Foch* regarding flexible defence, a problem with which the small and manoeuvre-oriented Reichswehr was preoc-cupied. For obvious reasons too, Liddell Hart wrote misleadingly in his *Memoirs* that they had discussed *Sherman*.[55] Later that year Walter von Reichenau, Blom-berg's chief of staff in East Prussia, wrote to Liddell Hart, informing him that he was translating *Foch*.[56] Colonel Sir Andrew Thorne, the British military attaché in Berlin from 1932–1935, recalled after the war:

> during that time there I could not fail to be impressed by the extent to which both Liddell Hart's and 'Boney' Fuller's books were being studied by officers of all ranks and arms in the German Army. I knew both Blomberg (Minister of War) and Reichenau (Chief of the Defence Staff) very well, and they were both engaged in translating books by these two authors for use for non-English speaking German officers. They frequently sought my help to elucidate some point which was not quite clear to them.[57]

As Guderian was to testify, the promotion of Blomberg and Reichenau following Hitler's rise to power had 'an immediate effect on my work. Both these generals favoured modern ideas, and so I now found considerable sympathy for the ideas of the armoured force, at least at the highest levels of the Wehrmacht.'[58]

It is time for a few words of summary on the influence of Fuller and Liddell Hart on the growth of the incipient German doctrine of armoured warfare during the second half of the 1920s. Although mostly prompted into action by tangible developments in British tank design, armour organization, and annual manoeuvres, the Reichswehr's primary access to these developments was through written, predominantly open sources. In the 1920s it was not even allowed to have a military attaché in Britain, but in contrast to Germany there was no difficulty in crystallizing a fairly complete picture of the British army from monitoring newspaper reports and parliamentary deliberations. It was therefore only to be expected that those in Britain who found greater expression in print were far more recognized in Germany than other, 'practical men', out of the limelight. This meant that Fuller and Liddell Hart enjoyed by far the highest profile, though other armour pioneers were also known as far as they published. Fuller was from the outset recognized as both the leading British armour practitioner and theorist, and from 1926 onward everything he wrote – even the most metaphysical and fanciful articles – was immediately seized upon in Germany. Liddell Hart, who in the mid-1920s had better access to print and who wrote prolifically and very lucidly, was slightly ahead in making his impact in Germany as a theorist, and his perceptive observations and ideas in the field of armoured warfare would always be regarded with great esteem in the Reichswehr. In addition, his columns in the *Daily Telegraph* remained a major source of information on all things military in Britain throughout the period. Until the mid-1930s Liddell Hart was the only permanent military correspondent in a British newspaper, the other newspapers employing special military correspondents mainly for the summer manoeuvres.

In conclusion, Fuller and Liddell Hart were highly influential in Germany through three main channels: directly, as theorists of modern mechanized war; indirectly, through the influence of their theories and actions on the development of British armour which from the mid-1920s was increasingly becoming a model for emulation for the Reichswehr; and as sources of information on British armour developments in the First World War and the interwar period.

Finally, how does Guderian fit into all this? As one historian has recently put it: 'Guderian is, at this point, just another general staff officer who supported . . . [the] new doctrine of mobile (tank) warfare as a means to overcome the impasse of the *Vernichtungsgedanke* in World War I.'[59] And even this assessment only really applies from the late 1920s. Guderian was posted to the inspectorate of transport troops as a captain in 1922 without any prior familiarity with the subject. The motorized units were predominantly concerned at the time with transportation, mainly of supplies, as was Guderian in his assignments with the troops and at headquarters. He took part in a number of trials and exercises in 1922–1924, only slowly developing an interest in the combat potential of motorization. In the first place, motorized transportation required protection against aircraft and also, when travelling through the battle zone, against ground troops. This, Guderian thought, could be provided by guns, armoured cars, and other mechanized combat troops. It was on these humble problems that he wrote his first short articles, published in *Militär-Wochenblatt* in 1924–1925. Tanks, forbidden to Germany, were still practically beyond his scope.[60]

Only in 1924 was Guderian entrusted with the responsibility for a series of exercises, 'intended to explore the possibilities of the employment of tanks, particularly for reconnaissance duties in connection with cavalry.'[61] Searching out solutions for the problems with which he was grappling, he began to explore the literature on mechanized warfare more widely, literature which was inevitably mostly foreign. He was introduced to it by Volckheim. 'The English and French had had far greater experience in the field and had written much more about it. I got hold of their books, and I learnt.'[62] The original paragraph Guderian would write in his memoirs concerning the influence of the writings of Fuller, Liddell Hart, and Martel, in that order, is unquestionably authentic.[63] For example, Guderian informed Liddell Hart that as far as he could recall he had first read his articles around 1923–1924.[64] After what we have seen above, there is no reason to doubt this evidence (the latter year being the correct one), especially as Guderian had been given no hints before by Liddell Hart regarding the publication dates of his earliest works on armoured warfare. Furthermore, at the time Liddell Hart's articles were prominently summarized in *Militär-Wochenblatt*, Guderian wrote his first articles for that journal and established close relations with its editor.[65]

From late 1924 to late 1928, the years German attention was turning to British armour theory and practice and the first steps were being taken to emulate them, Guderian was away from the inspectorate of transport troops and from involvement in decision-making. He was posted to the army's training branch and sent to the staff of an infantry division as instructor for tactics and military history.[66] By late 1928, he wrote, as 'I returned to my preoccupation with tanks . . . I was still lacking in all practical experience of tanks; at that time I had never seen the inside of one.' He derived practically everything he knew about them from a study of foreign literature.[67] By late 1927, reflecting on the direction German military thought had been taking, Guderian too arrived at the conclusion that the tank, aided by other mechanized arms and by aircraft, was the means for a renewal of the war of movement. Permanent mechanized formations were needed.[68] These were the ideas Blomberg, Heye, Fritsch, and others in the army's high command had already been officially endorsing and working for. Yet Guderian's memoirs only hint cursorily at some of this,[69] either because he was egocentric or because he did not think the 'prehistory' of the Panzer arm, before it was actually created in the 1930s, deserved more space.[70]

Indeed, things entered a new stage with the beginning of a new decade. The Germans were moving towards rearmament, a process finally set in motion with Hitler's rise to power. Earlier schemes were now beginning to materialize, including the creation of a German armoured force, which still looked up to the British as its model. Guderian was rising in rank and increasingly moving into positions of influence. His case really begins with the fact that in the autumn of 1931 he was nominated chief of staff to the new inspector of motorized troops, General Oswald Lutz. In this position Guderian became the main driving force of the inspectorate in the very years when rearmament and the expansion of the German army made the creation of Panzer divisions (and later corps) a real possibility. It hardly needs emphasizing that there was a wealth of difference between widespread awareness and even enthusiastic acceptance of the potential of such formations in theory, as shared by leading circles in the Reichswehr's high command in the second half of the 1920s, and the actual creation of the force with all the intricate problems and crucial decisions involved.

Towards armoured formations

Something of the developments behind the scenes was reflected by *Militär-Wochenblatt*. In the early 1930s the journal was publishing imaginary tactical problems on 'The Employment of an Independent Panzer Formation' (brigade size), and printing various schemes put forward in foreign professional periodicals for the structure of the armoured formation. As before, it was mainly to Britain that German eyes were turned.[71] A survey of 'The New Fast Mobile Formation' in European armies called it a British invention which was now being taken up by all countries.[72] Thus, concluded another article, 'An Independent Armoured Brigade in Battle': 'It is understandable that the rest of the world's armies pay much attention to the trials held in Britain.'[73]

The 1931 manoeuvres of the British armoured brigade under Broad's command were covered by *Militär-Wochenblatt*, which described them as 'the strongest concentration of tank formations since the war'.[74] Following the 1932 manoeuvres, British sources were cited to the effect that the range of which tanks were now capable was up to 240 km.[75] A survey of the manoeuvres, based on 'British journals', described the outflanking move of the armoured brigade against an infantry division. The latter was reported destroyed.[76] A large opening article with pictures, 'The Mechanization of the British Army', started with the words:

> The British army undoubtedly stands today ahead of all armies in the field of mechanization. . . . The frequent manoeuvres of the armoured formations at Salisbury are today the object of curiosity for all men of cultivation. They are in fact the highest school of mechanization.

The article reported that the British tank brigade was intended for independent raids behind the enemy's position, with the motorized infantry and cavalry moving in its wake to secure and hold the ground occupied by the tanks. The tank brigade was reckoned to advance in leaps of up to 150 km a day. Its operations would resemble those of sea warfare.[77] From 1932 Major Walter Nehring was first staff officer responsible for organization in the inspectorate of motorized troops (*Erster Generalstabsoffizier [Ia.] für Fragen der Organisation*) under Lutz and Guderian. His first little book on armoured warfare (1934) stressed the British tank brigade's employment for independent strategic (*operativ*) missions under a system of radio control.[78]

The surviving documents of the German general staff depict a similar picture from behind the scenes. A fully detailed, day-by-day survey of the armour manoeuvres of 1932, complete with maps and organization tables, was composed by the intelligence branch, signed by the chief of the general staff General Wilhelm Adam, and distributed in 250 copies.[79] The authors of the survey, who stressed the importance of the manoeuvres, complained that the reports available to them (not cited) still only allowed the most general picture. From 1933, however, this problem was remedied, for Germany now possessed a military attaché in London. The first to hold the post was Colonel, later Major-General, Baron Geyer von Schweppenburg. His political and military reports from Britain received the closest attention from, and were highly valued by, the Chief of the General Staff from late 1933, General Ludwig Beck.[80] One of his first dispatches to Berlin was a British official briefing of the tank brigade's manoeuvres of 1932.[81] Commenting on another of Geyer's early dispatches, an overall report on the British army, German intelligence concluded in respect to armour: 'General judgement: the British know that in their tanks they have a valuable weapon. There is an apparent tendency to rely on the effect of tanks against flanks and rear.'[82] In March 1934, in a new report on the state of the British army, Geyer wrote: 'In the mixed tank brigade the British army created the most important mobile "modern formation", which it holds to be necessary for powerful, long range, all-out strikes.' He further assessed that in creating a tank battalion in Egypt the foundation was laid for the creation of a second 'modern formation' – the light tank brigade.[83]

In October 1934 Geyer reported in full detail on the pioneering armour manoeuvres of that year in which the tank brigade and experimental Mobile Division under Percy Hobart and George Lindsay, respectively, for the first time tried out sweeping strategic penetrations.[84] A few days before, *Militär-Wochenblatt* had already printed a concise but highly informative survey of the manoeuvres. The tank brigade, it was written:

> used its high mobility to strike deep into the enemy's territory . . . a city was fixed as an important strategic target deep behind the enemy's front, where arms factories, a big railway station, and the enemy's main headquarters were located.[85]

Of curiosity interest, perhaps, at a later date the journal's exercise in translation from English read:

> The task given to the Mobile Force was to carry out a raid against various objectives – headquarters, ammunition depots, aerodromes, etc. – lying in an area of about 140 square miles. The raid entailed a penetration of some 60 miles into enemy country.[86]

In December the general staff issued its own conclusive support, signed by Chief of Staff Beck and again distributed in 250 copies.[87] I shall return to this later on.

Geyer was attending the British manoeuvres himself, and was regularly meeting with leading British politicians and military men. He too, however, derived a great deal of the information he passed on to Berlin from British newspaper reports, of which those by Liddell Hart in the *Daily Telegraph* were prominent. During the early 1930s articles and books by Fuller and Liddell Hart continued to be regularly printed in German professional journals, receiving the highest acclaim. Again, *Militär-Wochenblatt* was using extracts from their writings in its regularly featured exercises in translation from English.[88] Of the two, Fuller was universally recognized as the father of the new British school.[89] By now, however, he was going into retirement and, although remarkably alert and prolific, was withdrawing from active involvement with the Tank Corps.[90] By contrast, Liddell Hart was tightening his association with the tank brigade as it was beginning its most ambitious manoeuvres yet. Both as a writer and through his personal connections in the army and War Office, he was now constantly hammering his most cherished theme – the strategic, long-range use of armour.[91] Geyer could not have known of his close cooperation with Hobart during the mid-1930s, but he realized his special position all the same.

In October 1933 Geyer sent to Berlin an article by Liddell Hart in the *Daily Telegraph* on the deficiencies of the British army in modern equipment. In the covering letter he wrote: 'Following my report on Liddell Hart's role in the life of the British army (report no. 7, p. 3), I enclose a very informative article of his, which appeared after I had written my comments on the British army.' Geyer further wrote that Liddell Hart's well-known train of thought notwithstanding, the article could be taken as a mouthpiece for the younger generation in the British army. Unfortunately, Geyer's original report, dated 30 September 1933, including his original references to Liddell Hart and commented upon extensively by German intelligence, has apparently been lost.[92] In April 1934 Geyer reported on the creation of a support battalion in every infantry brigade in Britain. He again noted that Liddell Hart (name underlined), the source of the information, 'has a favoured relationship with the War Office'.[93]

As he gained experience in Britain, Geyer's wisely discriminating attitude to Liddell Hart's writings sharpened. In November 1934, sending to Berlin reports from the *Morning Post* on the reorganization of the British army, he praised the newspaper's military correspondent as very serious and highly regarded in the War Office. In this respect, he wrote, he was like Liddell Hart, except that he wrote in a sober and dry manner, not as a self-styled prophet of a doctrine.[94] It was, however, precisely in that aspect of his writings that Liddell Hart had no substitute. Thus, in April 1935, Geyer reported the publication of Liddell Hart's *When Britain Goes to War*.

> From the book I call the immediate attention of In.6 [the inspectorate of motorized troops] to the description of the last years' manoeuvres of the tank units (pages 223, 274, 281) [1931, 1932, and 1934 respectively]. A short discussion of the trials of the tanks in September 1934 [the experimental Mobile Division] is to be found on page 293.

Several days later Geyer replied to queries sent to him by In.6 in Berlin regarding the 1934 British armour manoeuvres (queries not in the file). In reply to one query he wrote that the answer could perhaps be found in page 293 of Liddell Hart's *When Britain Goes to War*. A week later he confirmed that the book had been sent to Berlin.[95] *When Britain Goes to War* (1935) is commonly assumed to be merely a revised edition of LH's *The British Way in Warfare* (1932); in fact, apart from the famous opening piece and a number of general essays on the theory of war, this is an entirely new book, incorporating many of LH's articles of the years 1931–1935 on the manoeuvres of the British armoured forces and constituting perhaps his most important book on the subject of armoured warfare. The comments of the German reviewer of the book in *Militär-Wochenblatt* are also worth quoting:

> Extraordinarily interesting, particularly for the leaders of our new Panzer formations, are the chapters in which the tactical trials over the years of the British tank brigades on the classical Aldershot training ground are described. The development of the experimental mechanized and armoured formations from 1918 to 1935 is especially emphasized. Like Zeus's sunlight and rain he bestows in describing the tactical missions in Aldershot love and blame on the leaders of the armoured formations.[96]

The reviewer concluded that 'the book should as quickly as possible be translated into German', which it was in the following year.

In the summer of 1935 the British manoeuvres again involved the tank brigade. In October Geyer sent his detailed report on the manoeuvres. He noted a growing opposition to the tank formation.[97] In December the intelligence branch of the German general staff composed its own even more comprehensive report, only partly relying on the attaché's report. Liddell Hart's articles in *The Times* featured here most prominently. As usual he needed no introduction. Among the many references to his reports and judgements were the following:

> Liddell Hart highly praises the fact that in these exercises the tanks were assigned a strategic [*operativ*] mission. In his view, it is against the nature of the armoured arm [*Panzerwaffe*] to be solely employed as a tactical assistant of infantry. According to Liddell Hart's opinion, the employment of the armoured brigade was not very fortunate. It should have been used further north in order to press more strongly on the enemy's rear communications. Besides, the danger that the brigade would get involved in a costly *tactical* battle would thereby have been diminished.

And again:

> According to Liddell Hart, the enemy's artillery is not the right target for the armoured brigade. The attack on artillery is too risky and requires an especially skilled battle technique from the tanks. . . . More successful would be an attack on the rear communications. Liddell Hart holds in the rest of his interpretation that armoured formations should be entirely *freed from their unarmoured baggage train*, or else the advantage of their mobility over great distances might be lost.[98] [emphasis in the original]

Although carefully collecting every piece of evidence testifying to his fame and influence, Liddell Hart was only generally aware at the time of this close attention to his writings, and would never have the specific details even later. In 1933 Sir Maurice Hankey, Secretary of the Committee of Imperial Defence, informed him: 'I heard from a high military authority in Berlin the other day that your and Fuller's writings are widely read and eagerly awaited throughout the German army.'[99] Liddell Hart and Fuller (who did not figure in Geyer's reports) dined with Geyer in 1934, but such correspondence as Liddell Hart and the German attaché had was formal. In July 1935 Geyer consulted Berlin on whether to invite Liddell Hart to attend the German manoeuvres (Fuller, a fascist and friendly to Nazi Germany, was invited, and attended). He wrote that *The Times* had the leading military reports and that it was now in general very pro-German, but added: 'I do not hold Liddell Hart himself to be truly friendly to Germany.' Invitations were issued, but for some reason Liddell Hart declined. When Geyer left London in 1937 he informed Liddell Hart that he had been given command of a Panzer division, but only after the war did he resume contact with Liddell Hart in a very friendly manner. He was the translator of Liddell Hart's *Memoirs* into German.[100]

The British model and the creation of the Panzer arm

The period discussed above was of course the one in which the German general staff and In.6, headed by Lutz and Guderian (from the spring of 1934 onward simultaneously also head and chief of staff, respectively, of the newly created command of the motorized troops; from late 1935, command of the Panzer troops), were working out and experimenting with the shape and role of the new Panzer units. The light interim models, Panzer I and II, began to come off the production line in 1934 and 1935, respectively, and the first three Panzer divisions were established in October 1935. Again the model for this development was clear. Colonel Sir Andrew Thorne, the British military attaché in Berlin between 1932 and 1935 – in whose house Guderian was a guest – told Liddell Hart in 1942 that the Germans had copied everything from Fuller and from him and that this had been common knowledge in Germany.[101] In 1958 Peter Paret recalled that it had been said in Berlin in the 1930s that Chief of the General Staff Beck wished six months could pass without him having to hear the name Liddell Hart.[102] But one does not have to rely solely on Liddell Hart's later records. As Thorne's words indicate, German sources of the time made no secret of whom they were trying to emulate. Guderian wrote in *Achtung Panzer!* (1937): 'After mature consideration it was decided that until we had accumulated sufficient experience on our own account, we should base ourselves principally on the British notions as expressed in *Provisional Instructions for Tank and Armoured Car Training* II 1927.'[103] In 1934 Major Walter Nehring, head of organization in the inspectorate of motorized troops, wrote programmatically that the future belonged to the British independent and strategic (*operativ*) use of armour in cooperation with aircraft.[104] And after a visit to Berlin in 1935 General Sir John Dill reported that he had been told by Lutz 'that the German tank corps had been modelled on the British.'[105]

Before continuing to trace the influence of British theory and practice on the development of German armour, we must first clarify another contentious historical question: who exactly in Germany deserves the credit for the creation of the Panzer arm? Hitler, War Minister Blomberg, the head of the ministerial office Reichenau, and the army's commander-in-chief, Fritsch, either supported or were sympathetic to the creation of large armoured formations. But by the German system of command the man whose responsibility it was to plan such things was the chief of the army's general staff, Beck. According to Guderian, he was conservative, in the mould of the old Moltke school, had no understanding of modern technology, and objected to the idea of Panzer formations larger than brigades or to the strategic (*operativ*) employment of armour.[106] Historians, however, have come to challenge this picture, in part or even in whole.

Beck has always been highly regarded as an intelligent and able chief of staff. Erich von Manstein, who as the head of the operations branch (*Abt. I*) from 1935 and deputy chief of the general staff (*Oberquartiermeister I*) in 1936–1938 was Beck's right-hand man and admired him deeply, very sensibly presented in his memoirs his chief's case (and his own) against Guderian's charges. Like other contemporaries, Manstein, who was in the best position to know, did not hesitate to describe Guderian as:

> the man who is rightly regarded the creator of the German Panzer arm. . . .
> Nobody familiar with the development of this question would dispute that without Guderian's tenacity and combative temperament, the German army would not have had the Panzer arm upon which its success in the first war years largely rested.

Manstein argued, however, that while Guderian saw only the Panzer arm, the general staff had to consider the entire army. He further argued that in contrast to the enthusiastic and impatient Guderian, the chief of the general staff could not responsibly commit the German army to revolutionary armour doctrine and organization before these had been thoroughly tested and proven in exercises and manoeuvres. Indeed, Manstein pointed out, it was after all Beck who, after sufficient evidence had been gathered, ordered the creation of the first three Panzer divisions and integrated them in strategic roles in operational planning and manoeuvres.[107] Manstein's arguments were revived, endorsed, and elaborated upon by historians once the relevant documents became available.[108]

All these are historically important points, which widen and balance the picture, but do not change it that much. Not unlike his British counterparts, Chief of the General Staff Beck was by no means inattentive and unreceptive to modern ideas, weapons, and techniques, but he was nonetheless far behind the cutting edge in the field. He was propelled forward to the length that he was only by the combined pressure of two forces, one from without and one from within. The former was developments abroad, especially in the British and French armies, which, as a highly professional staff officer who had the greatest respect for both armies, Beck studied most carefully. The latter was the abrasive and persistent Guderian, who the refined Beck personally disliked and in his constant hammering of his point shrewdly utilized the developments abroad to increase the pressure. The struggle went on until Beck's resignation in 1938, and although he implemented much of the radicals' programme, he came close to watering it down crucially throughout the period.

The German field service regulations, written by Beck in the early 1930s and renowned for their clarity of expression and style, are a good starting point for the development of his thought on armour. Part I (1933) is progressive in the 1918 mould, basically repeating the instructions of the previous manual (1923).[109] Tanks working in close cooperation with infantry and supported by artillery, mechanized engineers, and aircraft were to participate in the battle to break into the enemy's position with the aim of reaching its artillery zone. Part II of the manual (1934) emphasized that tanks ought to be employed at the decisive point of the battle, both in attack and defence. It repeated the instructions of Part I, but also added that tanks could be combined with other motorized troops into Panzer formations which might be employed in a more independent role against the enemy's flank or rear, or for a breakthrough. Motorized infantry and light armoured formations were also envisaged.[110] The two-sided pressure was beginning its work.

As Manstein simply put it: 'The general staff from early on envisaged the employment of large, independent Panzer formations, because of the pioneering British experiments and the writings of General Fuller and Captain Liddell Hart.'[111] We have already seen the reports composed in the general staff and those sent from London by Geyer von Schweppenburg, Beck's confidant, on the British armour manoeuvres of 1932 and 1934. Beck himself extensively studied the 1934 manoeuvres, and for a good reason. Following the creation of the British 1st Tank Brigade as a permanent formation in the spring of that year and its exercises in August in deep strategic penetration, an experimental Mobile Division had been tried out for the first time in September. The divisional exercise had been judged a failure, and Beck was therefore justified in being more reserved than some of his subordinates towards the idea of large armoured formations. But his careful analysis was unbiased, and while highlighting the problems encountered and the fact that even the widely experienced British army had found it difficult to handle large armoured formations, he left the question of their feasibility open for the time being. Furthermore, he assessed that the British would decide to proceed on their course and create a permanent Mobile Division.[112] The understandably cautious approach expressed in his remark to Guderian, 'But why should *we* be the first to create them [Panzer divisions]?', thereby lost much of its ground.[113]

Beck now subjected the use of tanks to a series of trials. In June 1935 a general staff tour played out (without troops) a German counter-offensive against a Czechoslovakian offensive in the Erz Mountains. The counter-offensive was to be carried out by infantry and three (still non-existent) Panzer divisions. Given the nature of the terrain chosen, it is not surprising that Beck concluded that the Panzer force had been able to operate effectively only in limited parts of the whole area and that it was 'the weapon of good opportunity'. He wrote that in the first stage of a campaign tanks must be employed in close cooperation with the infantry formations, and under their overall command, to break the enemy's position and reach his artillery zone. Only when this was achieved could Panzer formations be employed independently to exploit the success against the enemy's flank and rear or for the breakthrough. Beck therefore concluded that in addition to Panzer formations the army would need strong tank battalions and regiments, to be kept in the army's reserve for cooperation with the infantry in the main battle. This was to remain his position from then on. He also expressed reservations regarding the Panzer commanders' system of forward command.[114]

At the end of August 1935 the Panzer troops were assembled for trials as an experimental Panzer division. The trials focused on the basics of command, control, and tactical handling, and in this respect stood halfway between Broad's manoeuvres with the armoured brigade in 1931 and the British 1934 manoeuvres of the Mobile Division. The report by the command of the Panzer troops, signed by Lutz, stressed that the division was a purely offensive instrument both in attack and defence. It envisaged two types of mission for it: an independent strategic strike, launched by surprise immediately at the outbreak of war against the enemy's flank and rear; or cooperation with other formations. The report objected to the idea of giving each infantry division a tank battalion, or a regiment to each infantry corps. It maintained that this could only lead to local successes, and would disperse the armour. In what appears to be a fall-back position, it proposed instead that Panzer brigades would be created to support the infantry, while Panzer divisions would be concentrated in corps for independent strategic roles.[115]

Three Panzer divisions were established in October 1935, but for Beck this was only part of a general programme for the development of armour for the 'offensive army' now being created. He specified three roles for the tanks: support for the infantry attack ('infantry tanks'), fighting enemy tanks, and independent strategic employment in cooperation with other motorized arms. In addition to the Panzer divisions, of which he wanted no more, he planned to create a Panzer regiment for each regular corps by September 1939, which would be concentrated in Panzer brigades. This would give 36 battalions for cooperation with infantry as against 12 in the three Panzer divisions. Beck also planned the creation of semi-motorized and motorized infantry divisions and units, and mentioned the possibility of forming light mechanized divisions.[116] In short, Beck understandably looked around and wished to do what other leading armies were doing. At exactly that time, the British announced their intention gradually to motorize their infantry divisions and provide each with a battalion of infantry tanks for close support, in addition to the formation of the Mobile Division. These tank battalions would be concentrated in army tank brigades. And in the summer of 1935 the French conducted large-scale manoeuvres in Champagne to try out their new light mechanized division (DLM) and new motorized infantry divisions, established the year before.[117]

Beck repeatedly referred to the experience and conduct of other armies when defending his programme for a multiple role for armour against the general army office (Allgemeines Heeresamt), which suggested that nothing more than infantry tanks was required.[118] On the other side, his compromise solution was still very far from Guderian's insistence on avoiding any diversion of armour from the Panzer divisions. During 1936 the general staff under Beck's direction was planning the creation of four motorized infantry divisions, to be established in 1937, and of light mechanized divisions, to be formed one each year from 1937 onward. Both types, modelled on the French, were viewed principally as a strategic weapon for mobile use, and the former in particular was largely intended for cooperation with the Panzer divisions.[119]

Guderian, who would probably not have initiated the motorized infantry divisions himself, recognized their value. He totally objected, however, to the light divisions, which contained only one battalion of light tanks and were intended for 'cavalry-type' missions. He argued that strategic reconnaissance could better be carried out by aircraft, and that none of the missions specified necessitated a special Mobile Division.[120] The Polish campaign proved him right, after which the light divisions were converted to Panzer divisions. However, the main problem from Guderian's point of view was the Panzer units intended for infantry support. Until autumn 1938, after Beck's departure, no new Panzer divisions were created, and the number of tanks coming from new production and remaining out of these divisions was progressively outweighing those in the divisions. Two independent Panzer brigades were created, and more were planned. It has been claimed that Beck simply wanted to keep an open mind and all the options open until more practical experience was gained.[121] Indeed, ultimately, the German army did not find it difficult, in the wake of the Polish campaign, to concentrate all its armour in new Panzer divisions. Still, Beck's decisions cannot be explained away so easily. In the first place, his view regarding the multiple role of armour was principled and firm. Moreover, in 1936 he insisted on the production of a special, heavily armoured 'infantry tank', in addition to the main battle tank and fire-support tank (Panzer III and IV), ordered in 1935 and 1934 respectively.

A general staff document, issued by the organization branch but exactly repeating the views expressed by Guderian in an article he had written at that time, argued against Beck's request that in view of the strength of the modern anti-tank gun, heavily armoured infantry tanks would not be effective. The document suggested that it was better to concentrate all tanks in armoured divisions for strategic use. It also argued that because of industrial shortages the infantry tank would not be available in less than four to five years.[122] Nonetheless, by late 1936 Beck ordered the work on the infantry tank to continue.[123] The documentary record appears to break off at the end of that year, and it is therefore not exactly clear how the matter ended. Possibly the idea of infantry tanks was ultimately abandoned in favour of self-propelled artillery (*Sturmartillerie*), belonging to the artillery and intended for the infantry divisions. The first of these weapons were beginning to come off production in 1940. Their value, especially under conditions of scarcity in resources for armour and as they were intended for the un-motorized infantry, was hotly contested between Guderian and Manstein both before and after the war.[124] The point, however, is that had Beck had his way and the production of a large number of heavily armoured infantry tanks been taken up, this would have created the sort of divergence in tanks' mobility and characteristics that was to hinder the concentrated use of the Allies' numerically superior armour in 1940, and continued to plague the British army throughout the war. Just as, in his decision to create the first Panzer divisions in 1935, Beck was following on the British pioneering model, so he wanted to stay in line with developments abroad at the time that British tank policy itself was changing and was about to come closer to the French pattern with the British decision to drop the old universal 'medium' design in favour of the differentiated 'infantry' and 'cruiser' types.[125]

From all this it can be seen that, as in other countries, 'reactionaries' in regard to armour were a minority in the German army and general staff. By and large, the general staffs in practically all armies were dominated by 'mildly conservative' to 'progressive' officers who recognized very well the value of the tank and wanted to use it in large numbers for a variety of roles.[126] Beck too belonged to this category. Yet, however understandable the mainstream attitude of not putting all the eggs in one basket, the future belonged to the 'radical' programme which insisted on concentrating the tanks in armoured formations and opposed their being tied down to foot-walking infantry.

The significance of these diverging concepts should not be underrated. For the Germans, and probably also for the other powers, this was very much the source of the difference between triumph and disaster in 1939–1941. Later on in the war, when all armies learnt to handle and confront armour, tanks would again have to cooperate closely with the infantry for the break-in battle. But creating a special type of infantry tanks, attached to and designed to support un-motorized infantry, was never to prove a good idea.

Hence also the attitude of the creators of the German Panzer arm to Soviet armour and the doctrine of 'Deep Battle', of which the Germans possessed a comprehensive and accurate picture.[127] The conservative Beck gave little thought to the Soviets, even though the Red Army's armour theory and practice could be enlisted to support his views. He looked only to the two west European armies. But the younger officers of the Panzer arm were less prejudiced. Guderian, for example, like his friends, was fully and respectfully cognizant of the massive nature of Soviet Russia's industrialization and of the Red Army's mechanization. He visited the Soviet Union with Lutz in 1932, during the period of German-Soviet cooperation, and was much impressed by its tank production.[128] Like other German writers on the subject, he produced the data gathered by German intelligence and at the disposal of the command of the Panzer troops. With 10,000 tanks, 150,000 tractors, and 100,000 other vehicles, he wrote, 'the Red Army was ahead of all armies in mechanization, leaving Britain and France far behind.' He described the concept of 'Deep Battle' with its three armour group types, each with its special mission and special tank models: for close and long-range infantry support and long-distance penetrations.[129] However, this division of roles was exactly the thing he and his friends objected to. Nehring concluded that the permanent functional splitting of armour in that fashion was schematic and inflexible. Guderian wrote that while 'there is something to be said' for it, it 'demands a whole inventory of specialized tanks, with all the attendant disadvantages'. While he maintained that the Russians were leading in mechanization, Guderian held to a different model of armour doctrine and organization. He described three main variants in the world's armies regarding the employment of armour. In France the tanks belonged to the infantry and were also used in cavalry-type missions. 'Russia has advanced the furthest' in a second direction: using armour for infantry support as well as independently. However, 'the British since the war have taken a different route from the Russians and the French.' Their armour had been freed from cooperation with foot-walking infantry. And it was this direction that Guderian so strongly favoured.[130]

Although by then Guderian was, of course, well aware that the British themselves no longer strictly adhered to the model described, the 'British way' in the employment of armour had long become a symbol and the name of a generic type. Furthermore, for at least two years after the creation of the first Panzer divisions, Hobart's manoeuvres with the 1st Tank Brigade, in which long-range strategic penetrations were tried out, continued as before to be the focus of attention for the creators of the Panzer arm. It is important to realize that in those years the Germans, like the British in earlier years, were mainly preoccupied with the basics of organization, tactics, and, not least, the development of a workable system of radio communication, without which no far-flung operations could be seriously contemplated.[131] As mentioned before, the German inaugural manoeuvres of 1935 were of a rudimentary nature, and the armour enthusiasts were frustrated by the fact that no large-scale armoured manoeuvres took place in 1936. The first were held only in the autumn of 1937, when the 3rd Panzer Division and 1st Panzer Brigade of the 1st Panzer Division (800 tanks in all) took part in the great Wehrmacht manoeuvres across north Germany, in the presence of many distinguished foreign guests, including the much-impressed British Chief of the Imperial General Staff (CIGS), Field Marshal Sir Cyril Deverell.[132] It was only from that time on, not before, that the German Panzer force, still suffering from teething problems and lacking its planned war equipment (only the light Panzer I was in service), took a definite lead in the field of armoured warfare. From about that time it was also beginning to profit from its experience in the Spanish Civil War, especially in respect to the practicalities of air-ground cooperation.[133] Hitherto, in the years 1934–1936, the Germans had still been eagerly watching the British armour manoeuvres.

Manoeuvring deep into the enemy's rear

Unfortunately, the surviving official German records on the subject for the immediate prewar years appear to be particularly thin. For example, only one file, containing mainly political reports, seems to have survived from the dispatches sent by Geyer and by his successor from London in the years 1936–1938.[134] The intelligence files on the world's armies also provide a patchy picture.[135] Only the open publications give a better one. The British manoeuvres of 1936, including those of the Tank Brigade, were reviewed by *Militär-Wochenblatt*. It was reported that the brigade (200 tanks) exercised in tank-versus-tank warfare. Also, 'the marching exercises were nearly always so laid out that they would go far around the enemy's wing, and the armoured brigade would then strike at his flank.'[136] The

Militär-Wochenblatt's review of the brigade's manoeuvres in 1937 was based specifically on Liddell Hart's report in *The Times*. His emphasis on supplying the brigade from the air and off the land in which it was operating was cited.[137] Indeed, it was from this period that Liddell Hart would later receive a distant echo of the British and of his own influence on the development of German armour, which for years would remain his most significant one.

In 1941 Liddell Hart received a letter from the wife of a tank officer who in 1939 had been staying at the Bulgarian officers' club in Pleven. There she met Colonel Khandyeff, a Bulgarian officer who had been attached to a German armoured division a few years earlier. She described his recollection of his experience:

> The divisional commander was absolutely mad about the exploited and unexploited possibilities of tanks. His faith in armoured formations was such that he took a tremendous amount of pain in planting the same enthusiasm in the people under him. He spent his own money on providing copies of foreign books and periodicals, as well as on the services of a local tutor for the rough translations. His gods were General Fuller and Captain Liddell Hart. Liddell Hart, he considered, was the best analytical brain in the world, and his articles translated, read and studied, were discussed long before they would be vetted and sent from Berlin. While with him, Liddell Hart's accounts of the manoeuvres began to appear in *The Times*. As much as possible, every move of the manoeuvres was copied and put into practical demonstration. It was like a rehearsal of a play. The General was the happiest and busiest man, saying that Hobart gave him an answer to so many queries – and an inspiration. When a visiting anti-tank expert spoke of tank limitations as well as tank or no-tank country, quoting various opinions, including those of well-known people in England, the General impatiently dismissed him by saying – 'It is the old school, and already old history. I put my faith in Hobart, in the new man.'[138]

After the war, in 1945, General Wilhelm von Thoma, who had been one of the leading officers in the German Panzer arm since its inception, told Liddell Hart that he remembered the visit of the Bulgarian colonel and that he thought the general referred to was Guderian. He confirmed that the Germans had been keenly studying British military writings and had been closely watching the manoeuvres of the British armoured brigade in its successive forms. Guderian, when asked about the Bulgarian colonel's story by a fellow German general *before* Liddell Hart made direct contact with him, replied that he was not sure that it was his division that the Bulgarian colonel had referred to, but that it very well may have been: 'It is, however, certain that I had read many articles by Liddell Hart and that they were always of burning interest to me and that I have learnt much from them.' In response to Liddell Hart's direct enquiry in his first letter to him, Guderian clarified the point as follows: 'I don't remember him being present at the manoeuvres of my Panzer division. But as Thoma remembered meeting him and I was Thoma's divisional commander at the time, I think that Khandyeff was speaking about me in his letter.' After further urging by Liddell Hart he wrote: 'The account of Colonel

Khandyeff referred to 1935–36, when I was in command of the 2nd Panzer Division at Wurzburg.'[139] The later year was probably the correct one, first, because the German Panzer divisions were created only on 15 October 1935, whereas Hobart's exercises of that year had been held, as usual, in August; and second, because Guderian was promoted from colonel to major-general only on 1 August 1936.

This telling episode squares perfectly with what we have been seeing all along. Although in respond to later enquiries Guderian informed Liddell Hart that before the war he had read his *When Britain Goes to War*, *The Future of Infantry*, and *The Remaking of Modern* Armies,[140] it appears that it had been primarily Liddell Hart's *articles* that he had been following with 'burning interest' and from which he had been learning much.

In his book on Liddell Hart Mearsheimer incredibly missed the bulk of Liddell Hart's writings on armoured warfare, published chiefly in his columns in the *Daily Telegraph* and *Times*, rather than in his books and magazine articles. He therefore claimed erroneously that Liddell Hart had written very little about the theory of armoured warfare, had ceased almost entirely to write about and argue for it in the 1930s, and above all had never advocated deep strategic penetration before the Second World War.[141] It is now clear, however, that Liddell Hart's later comment, which Mearsheimer cites with ridicule as patently evasive, was in fact entirely true; in 1953 Liddell Hart wrote that the adoption of his ideas on the expanding torrent in tank doctrine, 'both here and in Germany . . . was produced not from any one exposition of it in a book, but rather by a constant reiteration of the keynote in my current articles commenting on manoeuvres etc.'[142]

All this also helps to explain another fact which has recently come to the attention of scholars and has greatly impressed them. Kenneth Macksey has pointed out that the bibliography to Guderian's *Achtung Panzer!* (1937) cites no work by Liddell Hart, even though he is mentioned in the text, whereas it does include books by Swinton, Fuller, Martel, and even de Gaulle. Macksey has interpreted this as proof that Liddell Hart's influence on Guderian was much more limited than the former boasted.[143] However, once we realize that Liddell Hart's main contribution to the theory of armoured warfare and to the creators of the Panzer arm was made in his articles, a fact of which at least Macksey seems to be aware,[144] the picture looks different. *Achtung Panzer!* was a collection of Guderian's lectures and articles, published to popularize the Panzer arm with the general public. The attached bibliography was in fact a small elementary source list (*Quellenverzeichnis*). More than half of the book was dedicated to tanks in the First World War, and so was about half of the entries in the source list. Only one book each was cited for Fuller, Swinton, and Martel, and in each case this book was in fact the record of their personal service with the tanks.[145] Liddell Hart had no comparable book, though had his *When Britain Goes to War* already existed in a German translation (translated the same year *Achtung Panzer!* was published), it probably would have been included in the source list. There was little point in referring readers to his scattered articles in the *Daily Telegraph* and *The Times*. Finally, there were the limits in which *Actung Panzer!* was written. The subtitle of the book was *The Development of Armoured Forces, Their Tactics and Operational* [*operativ*]

Potential, and in the text Guderian mentioned that 'the tank forces would gain by [the British way of employing them] not only a local, tactical importance on the battlefield, but one which extended into the operational sphere of the theatre of war as a whole.'[146] Still, as Macksey has noted, 'the latest thoughts about the ambitious role of Panzer divisions for deep penetration were muted' in the book.[147] And it was there that Liddell Hart's main contribution lay.

There is a related point which the episode of the Bulgarian colonel may illustrate. Guderian was using Liddell Hart's reports of the manoeuvres of the British tank brigade as a major source of information. He thought highly of him both as a theorist of armoured warfare and as a military critic. He was undoubtedly well aware of Liddell Hart's habit of using his reports to advance his own views. Nonetheless, he could not have known of the close cooperation, the virtual unanimity, and the stimulating exchange of ideas which existed between Liddell Hart and Hobart. In short, he could not have been fully aware of the degree to which Liddell Hart, through his partnership with Hobart during the period 1933–1937 in which deep penetrations by armoured formations were actually put into practice in Britain, was directly involved in, and contributing to, the advancement of this favourite idea of his. He went beyond Fuller's pioneering conceptions which had never progressed after having been last codified in *Lectures on FSR III* (1932).[148]

After the middle of the 1930s the fame of Fuller and Liddell Hart reached new heights in Germany. The first reason for this was that their books, and not only their articles, were now being translated and published, rather than being merely abstracted, serialized, and reviewed in German professional journals. The list of Liddell Hart's books translated before the war included: *The Future of Infantry* (1934), *Lawrence* (1935), *When Britain Goes to War (1937), Scipio* (1938), *Foch* (1938), and *The Defence of Britain* (1939). The equivalent list for Fuller included: *Generalship (1935), Memoirs (1939)*, and *The First of the League Wars (1939)*. There is little doubt that Fuller kept his senior place as the founding father of 'modern' armoured warfare and of the 'British school'. He was treated as such by all experts. He undoubtedly also held the senior place among Guderian's 'gods' throughout the interwar years. Fuller never actively sought accolades after the Second World War, but, as the British military attaché in Berlin in 1939, Brigadier T. Denis Daly, recalled: 'General Guderian . . . told me at length of his studies of the writings of Major-Gen. Fuller.'[149] In addition to everything else, Fuller was a fascist, and his political and politically charged writings were prominently cited in Germany. He visited Germany several times during the Nazi period as an official guest, was received with great honour, and met among others Hitler, Ribbentrop, and Hess.[150] Guderian's elder son has written in reply to Macksey's inquiries:

> As far as I know it was Fuller who made the most suggestions. Once before the war my father visited him. Fuller was certainly more competent as an active officer than Captain B.H. Liddell Hart. . . . At any rate my father often spoke of him while I cannot remember other names being mentioned at that time. . . . The greater emphasis upon Liddell Hart seems to have developed through contacts after the war.[151]

No matter how exactly this evidence is interpreted,[152] it was in fact mainly to Hobart, with Liddell Hart behind him, that Guderian's attention was turned in the years 1934–1937. For while Fuller remained the giant figurehead of armoured theory, armoured warfare during the mid-1930s was being rapidly developed by real fighting formations on the plains of southern England.

Another factor that greatly enhanced Liddell Hart's fame in Germany in the second half of the 1930s was the fact that he was now writing for *The Times*. Even more than before, he was now by far the most frequently cited foreign strategic authority – from Britain or from any other country – appearing in nearly every weekly issue of *Militär-Wochenblatt* over some years and always referred to with the greatest respect. Among others, his latest operational precepts were cited in the German journals practically every time they appeared in English. These forecast the growing strength of defence owing to the growing number of anti-tank weapons and the defender's growing ability to rush up mobile reserves to block the attacker's breakthroughs; but they also proposed methods which armoured formations might use under the changing conditions. Most importantly, Liddell Hart suggested where the best chances of armoured formations lay:

> One must use the rapidity of deep penetrating leverage to demoralize the enemy by creating repeated flanking threats which he would be unable to parry. One must advance on a wide front in order to be able to envelop the enemy and find and penetrate through gaps in his front. For that purpose it is best to advance on as many roads and with as many spearheads as possible, thus improving the chances of locating an 'inner flank'.

This is the method termed 'the expanding torrent', in which 'every subordinate commander should penetrate as deep as possible', and the reserves follow in order to support and exploit successes.[153]

Having already seen a great deal of the evidence for the decisiveness of British influence on the formation of the Panzer arm and the development of its characteristic doctrine, it is time to pause for a more general question, of a theoretical as well as practical significance. Liddell Hart's conception of the 'indirect approach' aimed (in Fuller's spirit) at disorienting, dislocating, and causing the disintegration of the enemy by a variety of means rather than the destruction of his armies in battle. Much of this was assimilated into British armour doctrine during the 1920s and 1930s. By contrast, was not the German Blitzkrieg, while employing revolutionary new weapons and techniques, in fact combat-oriented and aimed at great battles of encirclement and annihilation (*Kesselschlachten*), in line with traditional Prussian-German doctrine against which Liddell Hart had crystallized his ideas in the first place?[154] After all, Guderian's slogan was 'boot 'em, don't spatter 'em' [*Klotzen, nicht Kleckern*].[155]

To clarify this point one must turn to the famous concept of Blitzkrieg. It is an incredible and seemingly inexplicable historiographical fact that it has taken nearly half a century since the Second World War for the realization slowly to filter through that, contrary to widely held perceptions, Blitzkrieg was not the name of any official or at first even unofficial German doctrine allegedly crystallized during the interwar period.[156] It was a popular foreign designation, invented outside Germany following her first swift victories in the Second World War. Liddell Hart, who referred to the German 'Blitzkrieg' and 'lightning war' in 1939, even before the war, and later implied that he had invented the concept, was apparently not the only media man to have used the phrase. In July 1940 *Militär-Wochenblatt* wrote that the term Blitzkrieg 'stemmed from the ranks of our adversaries and quickly went into currency also with us'. Guderian wrote similarly in his memoirs: 'as a result of the successes of our rapid campaigns our enemies coined the word Blitzkrieg.'[157]

Being an undefined and retrospective concept, Blitzkrieg could be, and has been, given various and often arbitrary interpretations which regarded it as Germany's pre-planned strategy – politically, economically, and militarily. The fact that no such overall official doctrine existed does not, however, mean that what came to be known as Blitzkrieg at the beginning of the Second World War did not rest on some real developments and strategic ideas in prewar Germany. Militarily, traditional Prussian-German strategic doctrine (restated in the latest edition of the German field service regulations composed by Beck in 1933–1934) had always emphasized vigorous and offensive action in order to destroy the enemy's armed forces in great battles. The lessons of the First World War, while highlighting the need for total economic and social mobilization for a protracted war, also added new urgency to the traditional imperative of achieving a rapid decision, so that a war of attrition against a superior enemy coalition could be averted. As early as 28 February 1934, addressing army and Sturmabteilung leaders at the Reichswehr ministry, Hitler predicted that in order to gain living space in the east for the German people in the teeth of international opposition, 'short, decisive blows to the West and then to the East could be necessary'.[158] Both Hitler and the armed forces' high command largely promoted the air and mechanized forces as instruments for that purpose, and each of these forces had its own operational doctrine. Blitzkrieg evolved in stages in Poland and France, as all these factors unravelled and interacted in the political and strategic circumstances created during the first phase of the Second World War. It embodied many inherent tensions, reflecting Germany's fundamental material weaknesses as well as diverging operational concepts within her high command.

Let us examine the evolution of prewar German armour doctrine. From its inception the Panzer arm adopted a mode of operation which was influenced by the British theory that had served as its model. This mode was described as follows in a German news-sheet disseminated at the end of 1933 and dealing with the employment of the armoured formation:

> The manner of its engagement is not in prolonged battles but short, well-timed operations launched by brief orders. The principle is to use the battle tanks at the core of operations, to concentrate the main fighting force at the decisive point of action . . . on the principle of surprise in order to avoid or avert enemy defensive action.[159]

British armour doctrine, Lieutenant-Colonel Nehring of the Panzer arm wrote with approval, stressed rapid hammer-blows rather than drawn-out battles.[160]

Of course, Guderian and his friends wanted to employ the Panzer formations for achieving crushing military victories, sometimes by means of great encirclement battles. Yet they also had a more far-reaching conception regarding the manner by which decisive results could be brought about. As is well known, during the major Blitzkrieg campaigns in the West and in Russia intense disputes developed within the German high command regarding the pattern and aim of operations. The commanders of the infantry armies, more conservative in outlook and fearing, or struggling with, stiff enemy resistance along the fronts and flanks penetrated by the Panzer forces, wanted these forces to assist in ringing the enemy's armies in order to effect great *Kesselschlachten*. On the other side the Panzer leaders, most notably Guderian, resisted this request and saw the role of the mechanized forces as driving as deep and as fast as possible into the enemy's territory, with the view of bringing about a total collapse of his armed resistance and getting hold of his main communications and civilian centres.[161] Whether this was possible, especially in the Russian campaign, in which enemy resistance was much more stubborn than that which the Germans had experienced in the West and where spaces were many times vaster, is another matter. The idea, however, was that conducting or even winning 'ordinary' battles would only slow down the German offensive, cause the attrition of the Panzer force, and leave the enemy time to recover. Indeed, at least in the West in 1940, the most successful Blitzkrieg of them all, victory was gained 'strategically', with the Allies taken completely off balance, and hence with remarkably little hard fighting.

Moving to another, related question: both classical German doctrine and Liddell Hart were agreed in advocating calculated wide dispersion, changing into rapid concentrations; but was not Blitzkrieg in fact based on massive, knife-like concentrations of mechanized troops on narrow fronts? Critics of the conduct of British armour in the early North African campaigns have pointed out the tendency of the British commanders to spread their forces excessively widely, thus exposing them to the concentrated blows of the German Afrika Korps. Although Liddell Hart always insisted that dispersion must be calculated so as to enable effective cooperation, mutual support, and rapid concentration, it has been claimed with some justice that British armoured thinking during the 1930s, very much under his influence, emphasized very wide dispersion as the normal pattern of armoured operations.[162] Obviously, the balance between 'calculated dispersion' and 'rapid concentration' can be interpreted very differently in practice. Interestingly, in February 1942 *Militär-Wochenblatt* cited Liddell Hart's criticism in the *Daily Mail* of British handling of armour in North Africa. The problem, he wrote, was that the British used their armour piecemeal, whereas Rommel concentrated his.[163] Historically more significant, however, was the fact that German armour doctrine was itself similar to the British, from which indeed it had been derived.

The British armour's principle of advancing on a wide front, noted for example in Beck's study of the 1934 manoeuvres, was well recognized and was accepted without question by the creators of the Panzer arm.[164] Consider the most informative and programmatic of Guderian articles, published in 1939, where he wrote as follows: 'The success of an offensive will increase the wider the front and the greater the depth on which it is conducted.' He argued that breakthroughs on narrow fronts would especially expose the unarmoured tail of the armoured formations to enemy flanking fire. In addition, only on wide fronts would it be possible to employ several Panzer divisions together.[165] Well into the war in 1940 a German armour manual prescribed that decision should be attained by means of rapid concentration and deep penetration on wide fronts.[166] German and British conduct may have differed in reality during the war, but their respective doctrines were similar.

Naturally, there have been other claims of influence on the formation and doctrine of the Panzer divisions. Liddell Hart was deeply stung when Churchill told Parliament in 1942 that the Germans were implementing the ideas first formulated by de Gaulle, now leader of the Free French. For years Liddell Hart agonized about this rival claim, until he was able to solicit the support of the German generals against it. The evidence of the 1930s themselves show that de Gaulle's *Vers l'armée de métier* (1934) received considerable attention in Germany and was translated in 1935. The book and the author were prominently cited by the German armour pioneers, together with the names of other French advocates of mechanization like Generals Allehaud and Camon. All were presented in the propaganda campaign within the German army as the 'progressive' school in the French army.[167] Nonetheless, the veterans of the Panzer arm told Liddell Hart after the war that by the time de Gaulle's book had been published their own minds had long been made up, and that his proposals had anyhow been much too impressionistic and hazy to be of practical use.[168] And in view of the substance of de Gaulle's book and of what we know about the development of the idea of armoured warfare in Germany, this testimony seems flawless. Furthermore, de Gaulle, who dwarfed even Liddell Hart in egocentrism and vanity, made no reference in his book to his French predecessors, let alone, as Liddell Hart complained, to the British school. However, in 1943, in reply to a question about his book he said: 'But what about your best soldier, General Fuller? He was the prophet, we only followed him. . . . You will find prophesied in his books everything that the Germans did with tanks.'[169]

A more serious claim for influence can be made with regard to the Austrian reserve artillery general Ludwig Ritter von Eimannsberger. His book *Tank War* (1934; 2nd edn. 1938) started with a survey of the history of armour during and after the First World War. He contrasted the French and British armour philosophies, described the British pioneering manoeuvres from 1927 on and cited Fuller's ideas advocating tank fleets and an independent strategic employment of armour. Whether strategic penetration was possible, he wrote, was a question which in his own opinion still remained open; the British believed that it was.[170] Eimannsberger then outlined a prophetic blueprint for the use of armoured and motorized divisions, both in reserve for counter-offensives within the framework of defence in depth and for the attack.[171] Playing out a fictitious new Battle of Amiens, he projected a three-stage affair: first the enemy defensive lines would be penetrated in cooperation with armour; then the enemy's tank reserves would be defeated; and finally armoured and mechanized divisions would exploit the success independently.[172] Probably the most interesting parts of Eimannsberger's book were his proposals for the structure of the armoured and motorized divisions, which were fairly similar to those actually adopted by the German army in the following years. He also proposed tank brigades for the role of infantry close support.[173] It is difficult to know how influential Eimannsberger actually was. His book was widely known in professional circles in Germany, and after its publication he contributed regularly and prominently on armour to German military journals. Yet the officers of the Panzer arm denied that he had any significant influence, claiming that he merely reinforced already existing thoughts and plans.[174] Perhaps the nature of his proposals suggests some influence on Beck in the very period the latter was struggling to make up his mind on the question of mechanized forces. In any case, Eimannsberger dropped his doubts regarding the strategic use of armour only in the late 1930s and in the second edition of his book, and repeatedly pointed to Britain as his authoritative model.[175]

The structure of the Panzer division

Indeed, later divergences during the Second World War have obscured the degree to which the creators of the Panzer divisions in the mid-1930s based themselves on the British model in almost everything they did. The first German Panzer manoeuvres copied the tactical problems of the British exercises.[176] The British principle of mixing light and medium companies in the tank battalions was adopted by the Germans and was not abandoned before the end of the Second World War.[177] Moreover, the structure of the Panzer division itself was modelled on the British. Critics and historians of British armour have focused on the conflict between Lindsay and Hobart in 1934 regarding the structure of the Mobile Division, and have described the later development on the eve of the Second World War of the tank-heavy British armoured division, which contained very little infantry and not enough of the supporting arms. But for the creators of the Panzer arm in the mid-1930s all these were unknown and insignificant. What mattered for them was what was actually taking place at the time on Salisbury plain. All German reports of the 1934 manoeuvres of the experimental Mobile Division emphasized that this time the tank brigade (four battalions strong) cooperated within a single divisional structure with a motorized infantry brigade, motorized artillery, and motorized elements of reconnaissance, engineers, and signals.[178] This exactly was the structure adopted for the first Panzer divisions in 1935. The British continued to debate the structure of the armoured division throughout the second half of the 1930s, and the Germans kept track of the debate. But the British armoured division itself made its debut with little infantry and supporting arms only towards the beginning of the war, by which time the Germans no longer needed to look up to anybody.

Furthermore, even in theory there was little difference between the British and the Germans. Like the British theorists Fuller, Martel, and Liddell Hart, whom he cites in this context in *Achtung Panzer!*, Guderian from the late 1920s had come to advocate all-arm Panzer divisions, with all their elements armoured and mounted on cross-country vehicles. At the same time, like Fuller and his disciples, the exponents of armour in Germany emphasized the leading role of the tank in the division, and tended at first to relegate the infantry mainly to the second echelon of the division.[179] The tank, read a German news-sheet in 1933, 'is less well equipped to hold captured territory; for this purpose it is usually necessary to employ motorized infantry and artillery.'[180] Hence again the two separate brigade structures for the tanks and infantry in the Panzer divisions, which would be combined into makeshift 'battle groups' only later in the war. To be sure, in practice, from its first exercises the German Panzer arm embraced close inter-arm cooperation and did not restrict the motorized infantry of the Panzer divisions to defensive missions only.

Liddell Hart would later claim that in this the Germans were following his own teachings, and it is interesting to note that from the start summaries of his works in German journals did in fact consistently and clearly cite his concept of offensive armoured infantry or 'tank marines'.[181] His book *The Future of Infantry*, his first to be translated into German (1934), enjoyed great success in the Wehrmacht, though it must be said that it did not deal in any significant way with the infantry in the armoured division.[182] All the same, Liddell Hart's writings were far from being the decisive factor influencing German development in this respect. What mattered most was the Reichswehr's traditionally strong emphasis on inter-arm cooperation, highlighted in its service manuals, followed in practice, and taken up by the Panzer arm.[183] Liddell Hart's negative response to the inclusion of a whole motorized infantry brigade in the British Mobile Division of 1934 and his idea that the division needed only a handful of fully mechanized armoured infantry were noted by, but left no mark on, the Germans.[184] Unlike the British armour theorists, Guderian was grudgingly obliged to accept the fact that present resources and limited industrial capacity would not permit the provision of cross-country armoured vehicles to all the elements of the armoured division.[185] It was in all these, rather than in any principled difference of theory, that the reason for the later divergence between German and British armour lay.

Indeed, like all British armour pioneers, the creators of the Panzer arm held that the armoured division ought to be tank-heavy – the first Panzer division comprised 561 tanks. As in the British case, this number was almost halved by the time of the war, as medium tanks were replacing the earlier, lighter models.[186] However, as the number of tanks in the armoured division continued to decrease during the Second World War, both Guderian and Liddell Hart firmly objected to the change in their respective countries.[187] Furthermore, the former is on the record as enlisting the latter's authority in campaigning against the reduction at the highest levels of the Wehrmacht.

After the German victory in the West and in preparation for the invasion of Russia in 1941, the number of Panzer divisions was doubled from 10 to 20. However, low tank production forced the reduction of the tank establishment in each division from 220–320 to 150–200, whereas the other elements in the division remained unchanged, or had even been reinforced. The huge losses sustained in the war in the East further eroded the tank establishment of the Panzer divisions, and many of them comprised no more than a few scores of tanks. Some of the younger German field commanders thought the reduction in the number of tanks in the Panzer division in fact suited the changing conditions of warfare on the Eastern Front, where forces were now even, where spectacular tank drives were no longer feasible, and where closer cooperation between the tanks and the other elements in the Panzer division had become more necessary.[188] The British and Americans, who in 1942 imitated what the Germans had done out of expediency, would reach the same conclusion in their own theatres of war. Yet most of the armoured pioneers of the interwar period thought differently. When Guderian was nominated Inspector-General of Armoured Troops after Stalingrad, one of his main objectives was to bring the tank strength of the Panzer division back to its 1938 level of 400 tanks, even if that meant a reduction in the overall number of armoured divisions. To be battle-effective, he explained in a meeting with Hitler and senior staff officers on 9 March 1943, the Panzer divisions had to be very strong in tanks. To support his point, Guderian read out to Hitler and the attending officers a recently published article by Liddell Hart on the organization of armoured forces – past and present.[189]

This incident, not cited by historians, illustrates the point that Guderian was keeping track of Liddell Hart's writings even during the war. Indeed, it will be remembered that the most solid piece of evidence for the Liddell Hart-Rommel 'connection' is also a specific comment by Rommel in the spring of 1942 on an article Liddell Hart had written on the recent campaign in the Western Desert.[190] Monitoring open enemy sources is one of the most elementary functions of any intelligence service, and German intelligence was obviously distributing the material to interested parties in the German army. Furthermore, not only German confidential reports but also the open *Militär-Wochenblatt* continued its years-old habit of printing what the two famous British military critics were writing. In 1941 Liddell Hart, who had left *The Times* in 1939, again began to write regularly on the war for the *Daily Mail*, and Fuller also was writing for a number of newspapers. From summer 1941 their commentaries on the war, especially Liddell Hart's, were cited weekly in *Militär-Wochenblatt*. The journal ran a regular section, 'Wehrpolitische Rundschau', where Liddell Hart and Fuller figured prominently as virtually the sole enemy writers to be cited side by side with Churchill, Eden, and other British cabinet ministers. As before, they remained household names in Germany. Of course, there was now an additional reason for their popularity in that country: both held pessimistic views about Britain's prospects of 'winning the war' against Germany. Thus by late 1942, as it was becoming clearer that the war was not going in Germany's favour, citations in *Militär-Wochenblatt* from foreign journals and specifically from Liddell Hart and Fuller began to dwindle.

Conclusion

Leaving behind the conjectures and extrapolations from Liddell Hart's disputed papers and turning to the interwar German records themselves, the results produced may appear either exciting or dull. The traditional view of a decisive British influence on the formation of the Panzer arm returns with a vengeance, with some important modifications but on the whole greatly strengthened by often tantalizing documentation. True, Liddell Hart pressurized the German generals after the Second World War and then manipulated their evidence. But great as the German generals' debt to him at that time may have been, the idea that he could bring them to tell his desired story without the strongest foundation in reality should have been recognized on the face of it as incredible. Where they did not want to cooperate, they did not; and where they did, it was because what Liddell Hart wanted them to say was not far from what they knew to be true. From his point of view, Liddell Hart was on the whole working to get the credit he knew was his due. He undoubtedly focused the whole story around himself, mainly at Fuller's expense. In addition, the paramount influence of the practical development of the pioneering British armoured forces, in the context of which the influence of Fuller and Liddell Hart is to be understood, may not have been as clear to all those nourished on his version of events as it was to Liddell Hart himself. Still, the growth of the belief that that version was fundamentally false, for which Liddell Hart's own mischief is largely responsible, must be seen as a sobering lesson, if one is still needed, about the way historical myths may take root even in respect to relatively recent times and even in the scholarly community.

Notes

1 K. Macksey, 'Liddell Hart: The Captain Who Taught Generals,' *Listener*, Dec. 28, 1972, p. 895; idem., *Guderian: Panzer General* (London, 1975), pp. 40–1; idem., *The Tank Pioneers* (London, 1981), pp. 118, 216; elaborated and expanded by J. Mearsheimer, *Liddell Hart and the Weight of History* [hereafter: *LH*] (London, 1988), pp. 160–7, 184–201. For LH's immense trouble on the German generals' behalf, see his papers, deposited in the Liddell Hart Centre for Military Archives, King's College, London, esp. 9/24; I am grateful to the trustees of the Centre for granting me permission to quote from the documents to which they hold copyright. See also B. Bond, *Liddell Hart: A Study of His Military Thought* [hereafter: *LH*] (London, 1977), pp. 180–8, 227–8.
2 Manstein files 9/24/71 and 9/24/124; Paget to LH, 12 Apr. 1951, LH to Paget, 1 May 1951, in 1/563; R. Paget, *Manstein: His Campaigns and His Trials* (London, 1951), p. 22; B.H. Liddell Hart, *Memoirs* (London, 1965), pp. ii, 203–4; Mearsheimer, *LH*, pp. 188–9; Bond, *LH*, pp. 232–3.
3 See 9/24/24, 9/24/50 and 1/776; *The Rommel Papers*, ed. B.H. Liddell Hart (London, 1953), pp. 203, 299, 520; Mearsheimer, *LH*, pp. 191–201.
4 LH to Guderian, 12 Mar., 25 Feb., and 30 Apr. 1949; Guderian to LH, 2 Mar. and 30 July 1949, in 9/24/62.
5 6 May 1950, in 9/24/62.
6 Heinz Guderian, *Erinnerungen eines Soldaten* (Heidelberg, 1951), p. 15; LH to Guderian, 6 Apr. 1951; Guderian to LH, 23 Apr. 1951, both in 9/24/62; Heinz Guderian, *Panzer Leader* (London, 1952), p. 20; Mearsheimer, *LH*, pp. 164–5, 189–90.
7 Macksey, *Guderian*, p. 41; idem., *The Tank Pioneers*, p. 118; Mearsheimer, *LH*, pp. 190–1.
8 Walter Nehring's useful *Die Geschichte der deutschen Panzerwaffe, 1916 bis 1945* (Berlin, 1969) was also written by a veteran of the Panzer arm and Guderian's subordinate.
9 J.S. Corum, *The Roots of Blitzkrieg, Hans von Seeckt and German Military Reform* (Lawrence, KS, 1992).
10 *Heeresdienstvorschrift 487, Führung und Gefecht der verbundenen Waffen* (Berlin, 1923), esp. paras. 347, 362, 523–37, 551. Similarly, see the guidelines for an exercise

prepared by the Inspecteur der Verkehrstruppen (In.6[K]), 1 July 1924, in RH 39/115 (henceforth all German documentary references are to the Bundesarchiv/Militärarchiv, Freiburg im Breisgau). See also Corum, *Seeckt and German Military Reform*, pp. 122–6.

11   E. Volckheim, *Die deutschen Kampfwagen im Weltkrieg* (Berlin, 1923); idem., *Der Kampfwagen in der heutigen Kriegsführung* (Berlin, 1924); for his articles in *Militär-Wochenblatt* [hereafter: *M-W*], see 25 July 1924, p. 718; 4 Aug. 1924, pp. 119–22; 11 Oct. 1925, pp. 465–8; 18 Apr. 1926, pp. 1409–12; he was also the chief writer in six special supplementary issues entitled *Der Kampfwagen* 'The Tank'), published monthly between Oct. 1924 and Mar. 1925; an example of his work from Aug. 1924 on tank instruction in the army can be found in RH 12–2/51; see also Corum, *Seeckt and German Military Reform*, pp. 126–30.

12   F. Heigl, *Taschenbuch der Tanks* (Munich, 1926); the 3rd edn. (1935, 1938) expanded into three vols; Heigl's contributions to *M-W* can be found in nearly every other issue; for his consultation to the Reichswehr, see e.g. 14 Oct. 1924, in RH 39/115. See also Corum, *Seeckt and German Military Reform*, p. 103.

13   For the Heavy and Light Tractors, see W. Spielberger, *Die Motorisierung der deutschen Reichswehr, 1920–1935* (Stuttgart, 1979), pp. 281–350; W. Oswald, *Kraftfahrzeuge und Panzer der Reichswehr, Wehrmacht und Bundeswehr* (Stuttgart, 1982), pp. 340–6; Corum, *Seeckt and German Military Reform*, pp. 112–18. On the cooperation with the Soviet Union, see M. Zeidler, *Reichswehr und Rote Armee, 1920–1933* (Munich, 1993), pp. 188–98.

14   E. Volckheim, *Betrachtungen über Kampfwagen-Organisation und -Verwendung. Zu einer Abhandlung des englischen Majors Sherbrooke* (Berlin, 1924), p. 1; idem., *Kampfwagen und Abwehr dagegen* (Berlin, 1925), pp. 3, 9–10; both booklets repr. from *Wissen und Wehr*; Heigl, *Taschenbuch der Tanks*, pp. 78, 322–6.

15   *M-W*, 25 July 1924, pp. 713–15; the article had already been briefly summarized when the relevant issue of the British journal had been routinely reviewed: *M-W*, 5 May 1924, p. 578; for the second piece, see *M-W*, 11 Nov. 1924, p. 501.

16   For LH's adoption of Fuller's ideas on armour – and his later development – see Azar Gat, 'Liddell Hart's Theory of Armoured Warfare: Revising the Revisionists,' *Journal of Strategic Studies*, 19:1 (Mar. 1996), pp. 1–30.

17   'Gedanken über eine allmähliche Mechanisierung in der englischen Armee,' *M-W*, 18 Dec. 1924, pp. 649–51 (signed '21').

18   The *M-W* cuttings of 5 May, 25 July, and 11 Nov. 1924 can be found in LH's archive: 7/1919/13. As early as 10 July 1925, long before anyone could anticipate the German future exploits, the *Daily Telegraph*'s announcement of LH's appointment as its military correspondent already mentioned among his achievements that 'The German Ministry of War translated for circulation to their army his scheme for the progressive mechanization of the army'; also see LH's record of achievements of 1930, 11/1930/41.

19   T3, 'Bemerkungen zu den englischen Manövern 1924,' 1 Dec. 1924, esp. pp. 4–5, 10–11, in RH 2/1603; other intelligence surveys of the British army are in the same file. See also Corum, *Seeckt and German Military Reform*, p. 132.

20   *M-W*, 11 Jan. 1925, pp. 761–3 (signed '21').

21   For the Vickers Medium and the manoeuvres, see e.g. *M-W*, 25 May 1925 (Heigl's), the speed of the tank is here put at 40–45 km/h; 18 Sep. 1925, p. 382; 4 Oct. 1925, p. 449; 25 Oct. 1925, p. 559; 4 Nov. 1925, p. 587; 4 Dec. 1925, pp. 777, 784; 11 Mar. 1926, p. 1215; 25 Mar. 1926, p. 1287; 18 Apr. 1926, p. 1414. *Kriegs und Militärorganisationische Gedanken und Nachrichten aus dem Auslande* [hereafter: *KGA*] II (1926), 5:13–14, also summarizes the 1925 manoeuvres, stressing their futuristic character.

22   *M-W*, 11 Dec. 1925, pp. 771–2 (signed 'Mügge').

23   *M-W*, 18 July 1926, p. 87 (signed 'Mügge'); Fuller is cited in another article on the following page.

24 *M-W*, 11 Jan. 1926, pp. 913–18 (signed '83'); the author expanded these themes in a three-part article: *M-W*, p. 18 and 25 May, and 4 June 1926, pp. 1559–62, 1591–5, 1628–30.

25 For *Paris*, see *M-W*, 4 Sept. 1925, p. 291; and the general staffs foreign literature journal *KGA* II (1926), 8:23–6; see also e.g. LH's reports on Martel's one-man tank: *M-W*, 11 Sep. 1925 and 25 Jan. 1926, pp. 472, 1007; *KGA* II (1926), 16:17–18; LH's articles in the *Daily Telegraph* on a variety of issues were often cited in the journal.

26 *M-W*, 11 Oct. 1925, pp. 469–72.

27 For some early references to Fuller, apart from those already cited, see *M-W*, 15 Apr. 1924, p. 531; 11 Oct. 1925, p. 473. *KGA* II (1926), 2:21–8, summarizes his ideas as described in the *Daily Telegraph* (by LH) where he was named only as 'a high-ranking British Officer'; also see *KGA*, 6:10. On his official nomination, background, and views: *M-W*, 25 Feb. 1926, p. 1157.

28 *M-W*, 11 Aug. 1926, pp. 201–2; *KGA* II (1926), 9:5–31; 10:5–29; 11:14–31; both journals emphasized his official nomination and expressed the expectation that he would have a decisive influence on the future development of the British army. An article by J.F.C. Fuller also opened the collection *Kampfwagen und Heeresmotorisierung* (Berlin, 1926): *M-W*, 11 Aug. 1926, p. 214.

29 Copies (n.d.) in RH *8/*v. 1745 and 1939.

30 This is rightly pointed out by Corum, *Seeckt and German Military Reform*, esp. pp. 103, 131, 142; however, influenced by the recent revelations regarding LH's manipulations of the evidence on the subject, Corum wholly underrated his influence on the Germans (pp. 141–2), as the following will attempt to show.

31 29 May 1926, in RH 2/2195; *M-W*, 4 Aug. 1926, p. 146; the German 'englisch' and 'England' will be regularly translated as 'British' and 'Britain'. See also Corum, *Seeckt and German Military Reform*, p. 132.

32 *M-W*, 25 Oct. 1926, pp. 553–5. *The Daily Telegraph* (LH) is often cited, e.g. on tank fire on the move: *M-W*, 18 Nov. 1926 (signed '83'); the same in *KGA* II (1926), 14:21–7.

33 10 Nov. 1926, in RH 39/115.

34 Cited by M. Geyer, 'German Strategy in the Age of Machine Warfare, 1914–1945,' in P. Paret (ed.), *Makers of Modern Strategy* (Princeton, NJ: Princeton UP, 1986), p. 559; Geyer too emphasizes the British influence in these years.

35 *M-W*, 11 Apr. 1927 (signed '99').

36 Announcements on the beginning of the manoeuvres appeared in *M-W*, 11 and 18 Sep. 1927, pp. 353, 384; the report itself appeared on 11, 18 and 25 Oct. and 4 Nov. 1927, pp. 501–7, 540–3, 568–71, 607–8 (signed '96'); this time it was mainly based on *The Times* rather than on the *Daily Telegraph*. Interestingly, however, a translation of a *Sunday Times* article on the manoeuvres in *KGA* IV (1928), 1:5–9, reads (p. 7; retranslated into English): 'As Captain Liddell Hart, perhaps our best military writer of the new school, writes in his new book' – a note in the German journal explains: *The Remaking of Modern Armies*. For the Inspectorate of Transport Troops' report, see In.6 (K), 17 July 1928, in RH 39/115.

37 *KGA* IV (1928), 3:13–17; *M-W*, 25 Dec. 1927, pp. 909–10. Also in the same issue, pp. 893–4, on the Mechanized Force and criticism of Fuller in Britain as 'fanatic'; the report opens with the sentence: 'Britain is known as the leading country in the mechanization of the army.'

38 *M-W*, 25 Feb. 1928, pp. 1219–20 (signed '96'); Guderian, *Panzer Leader*, p. 22; Heinz Guderian, *Achtung Panzer! The Development of Armoured Forces, Their Tactics and Operational Potential* (London, 1992; German original 1937), pp. 167–8; idem., *Die Panzertruppen und ihr Zusammenwirken mit den anderen Waffen* (Berlin, 1937), pp. 15–16. See also Macksey, *Guderian*, pp. 48–9.

39 *M-W*, 4 Oct. 1928, pp. 495–6.

40 *KGA* III (1927), 3:18–23; IV (1928), 13:5–25,14:5–28; also 15:5–10; *M-W*, 25 Jan. 1928, pp. 1048–52.

41 *KGA* IV (1928), 23:6–9; V (1929), 17:4–13, 21:5–17; VI (1930), 1:19–22, 2:12–19, 4:4–16, 8:5–19; 10:12–16; 16:5–12.

42 For Fuller, on top of what has already been cited, see e.g. *M-W*, 4 July 1927, p. 11; 11 Jan. 1928, pp. 966–70; *KGA* III (1927), 5:5–21; IV (1928), 15:16–22; V (1929), 3:1823; 13:7–13; 15:5–15; VI (1930), 14: 24–26. For LH, see e.g. *M-W*, 11 Apr. 1927, p. 1412; 4 Nov. 1928, p. 675; 18 Jan. 1929; *KGA* III (1927), 6:20–1; and numerous reports from the *Daily Telegraph* on the British army.

43 In addition to books already cited, see the reviews in *M-W* of Liddell Hart's *Scipio*, 25 Jan. 1927, pp. 1027–8; *Great Captains Unveiled*, 4 Dec. 1927; *The Decisive Wars of History*, 18 Dec. 1929, pp. 905–6; *Sherman*, 4 Apr. 1930, p. 1466.

44 For criticisms of Fuller's *The Reformation of War* and *On Future Warfare*, see e.g. *M-W*, 11 Sept. 1926, pp. 321–3 (signed '12'); the consecutive 'Is Fuller Right?' *M-W*, 4 Dec. 1929, pp. 808–9 (signed '21'), 4 Jan. 1930, pp. 975–6 (signed '97'), 11 Jan. 1930, pp. 1010–2 (signed '139'), all sensibly qualifying rather than rejecting. Victor Germains's critical *The 'Mechanization' of War* (London, 1927) was extensively summarized in three issues of *KGA* IV (1928), 16:5–28, 17:5–31, 18:5–18.

45 See e.g. *M-W*, 11 Apr. 1927, p. 1412; 25 June 1928, pp. 1893–5; 18 Sept. 1929, pp. 401–3; 11 July 1930, pp. 1816–17; *KGA* V (1929), 8:11–16. Parts of Ernest Swinton's, *Eyewitness* (London, 1933), on the origins of the tank, were translated by German intelligence and circulated in typescript: RH 8/v. 1936–7; see also RH 8/v. 1935.

46 See *Wehrgedanken des Auslandes* (the journal's new title from 1931 on), XIV (1934).

47 Quotations respectively from *M-W*, 18 July 1926, p. 87; 25 Jan. 1927, p. 1027; 4 Nov. 1928, p. 675.

48 11/1948/38; Bond, *LH*, p. 234.

49 Geyer, 'German Strategy,' p. 559; Nehring, *Panzerwaffe*, pp. 54–6.

50 In.6 (K) (signed 'Lutz'), 1 and 18 June 1929, in RH 39/115.

51 T4 (signed 'Blomberg'), 1 Sep. 1929, in RH 39/115; see also Geyer, 'German Strategy,' p. 559.

52 R. Ogorkiewicz, *Armoured Forces* (London, 1970), pp. 17–18, 87. The American copying of the British manoeuvres (and the influence of Fuller and Liddell Hart) is revealed on the basis of the documents by John Hendrix, 'The Interwar Army and Mechanization: The American Approach,' *Journal of Strategic Studies*, 16 (1993), pp. 77–81.

53 See the official history, based on the documents: L. Ceva and A. Curami, *La Meccanizzazione dell'esercito italiano dalle origini al 1943* (Rome, 1989), pp. 113–32.

54 J. Erickson, *The Soviet High Command* (London, 1962), pp. 263–70; R. Simpkin, *Deep Battle: The Brainchild of Marshal Tukhachevski* (London, 1987), pp. 38 and passim.

55 Cf. LH, *Memoirs*, I. pp. 171–2, with his records at the time: 7 and 8 Mar. 1932, in 11/1932/1 and 11/1932/9; this has been pointed out by Mearsheimer, *LH*, pp. 162–3.

56 Reichenau to LH, 28 Nov. 1932, 9/24/87/R.

57 Thorne to Hankey, 22 Mar. 1946, 13/45; Bond, *LH*, p. 219.

58 Guderian, *Panzer Leader*, pp. 29, 37, 48.

59 Geyer, 'German Strategy,' p. 558.

60 Guderian, *Panzer Leader*, pp. 19–22; 'Truppen auf Kraftwagen und Fliegerabwehr,' *M-W*, 25 Sept. 1924, pp. 305–6; 'Straßenpanzerkraftwagen und ihre Abwehr,' *Der Kampfwagen*, I, Oct. 1924, pp. 5–8, 'Aufklärung und Sicherung bei Kraftwagenmärchen,' *op.cit.*, II, Nov. 1924, pp. 13–16; 'Die Lebensader Verduns,' *op.cit.*, IV, Jan. 1925, pp. 28–31, 'Kavallerie und Straßenpanzerkraftwagen,' *op.cit.*, V, Feb. 1925, pp. 37–8. The backwardness and esoteric nature of these contributions have been pointed out by Corum, *Seeckt and German Military Reform*, p. 139.

61 Guderian, *Panzer Leader*, p. 21.

62 Ibid., p. 20.
63 Ibid.; however, it probably summarizes his development over a number of years from 1924 on. Martel, for example, was occasionally cited in German periodicals in the second half of the 1920s, especially in connection with his one-man tank, but it was only with the publication of his *In the Wake of the Tank* (1931; German trans. the same year) that his full significance for the development of armour theory as early as the First World War became clear.
64 Guderian to LH, 19 Mar. 1949, 9/24/62.
65 Guderian, *Panzer Leader*, p. 20.
66 Ibid., pp. 21–2.
67 Ibid., p. 22; he first saw a real tank either during a visit to Sweden in 1929 (p. 23) or, perhaps, in connection with the clandestine German production.
68 Major Heinz Guderian, 'Bewegliche Truppenkörper. Eine kriegsgeschichtliche Studie,' *M-W*, 11, 18, 25 Nov., 4, 11 Dec. 1927, pp. 649–53, 687–94, 728–31, 772–6, 819–22. See also Bradley, *Guderian*, p. 166. For some unknown reason and against the printed evidence, Guderian's, *Panzer Leader*, p. 24 (mentioning the influence of the British exercises), date this decisive step in his development to 1929.
69 Ibid., pp. 29, 31.
70 Sharp criticism of Guderian's egocentrism is expressed by Corum, *Seeckt and German Military Reform*, esp. pp. 137–9; Corum, however, overstates a good case not only by unconvincingly claiming for Seeckt the crown of military innovativeness but also by blurring the difference between the Reichswehr's early attention to tanks, which went along conventional lines, and the new ideas developed from the mid-1920s under British influence. Guderian's cursory references to Volckheim and Heigl in *Panzer Leader*, pp. 20–1, is in this respect understandable. On this, see the balanced judgement of S.J. Lewis, *Forgotten Legions: German Army Infantry Policy, 1918–1941* (New York, 1985), p. 18.
71 See the tactical problem in *M-W*, 18 and 25 Jan. 1931, pp. 1200–4, 1237–43 (signed 'Lieutenant Wedel'); for various (unofficial) British proposals for the organization of an armoured division, see *M-W*, 11 July 1930, pp. 1816–17, 25 June 1931, pp. 1893–4. Also relevant are *M-W*, 25 Apr. 1931, pp. 1561–4; 25 Oct. 1933, pp. 509–11; 25 Sept. and 4 Oct. 1931, pp. 433–9, 469–74; 25 June 1934, pp. 1659–62.
72 *M-W*, 11 and 18 Dec. 1932, pp. 721–4, 756–61 (signed 'Lieutenant Faber du Faur').
73 *M-W*, 4 May 1933, pp. 1340–3 (signed '349').
74 *M-W*, 25 July, 11 Aug., and 11 Nov. 1931, pp. 139, 223, 661–2; quotation from p. 661.
75 *M-W*, 4 May 1933, pp. 1340–3 (signed '349').
76 *M-W*, 25 June 1933, pp. 1566–7 (signed 'Major-General Zölss').
77 *M-W*, 11 Apr. 1934, pp. 1259–62 (signed 'M. Braun').
78 Major Walter Nehring, *Kampfwagen an die Front! Geschichte und neuzeitliche Entwicklung desKampfwagens ('Tanks') im Auslande* (Leipzig, 1934), pp. 20–1; there is no need to dwell on the difficulty of translating the German *operativ*, standing in an intermediate position between the English 'strategy' and 'tactics' and signifying combat strategy in the theatre of operations.
79 'England. Die Manöver der Kampfwagentruppen, Sommer 1932,' 32 pp., 16 May 1933, copies in RH 2/2968 and RHD 18/137. Much of the material for the years 1932–4, contained in RH 2 and RH 12, is reportedly missing; but see the file by T3, 'Motorization and Mechanization in Britain at the Beginning of 1931,' 15 pp. plus pictures, 10 Feb. and 20 Mar. 1931, RH 12–6/v.22.
80 His memoirs, Freiherr Geyer von Schweppenburg, *Erinnerungen eines Militärattachés, London, 1933–1937* (Stuttgart, 1949), focus mainly on the political-strategic aspect of his mission.
81 4 May 1933, RH 2/1881; all reports sent to T3. On 20 July 1933, Geyer reported that there would be no concentration of the armour units that year.
82 T3, *op.cit.*, 26 Oct. 1933; Geyer's report itself (30 Sept. 1933) is apparently missing.

83 19 Mar. 1934, in RH 2/1882, quotation from p. 8.
84 10 Oct. 1934, in RH 2/1882, 15 pp. plus many sketches and appendices.
85 *M-W*, 4 Oct. 1934, pp. 489–94, quotation from p. 492 (signed '366'); also *M-W*, 25 Oct. 1934, pp. 610–16.
86 *M-W*, 25 Dec. 1935, p. 1056 (source not cited).
87 'England: Manöver des Panzerverbandes 18 bis 21.9.1934,' Dec. 1934, 36 pp. plus appendices and sketches, copies in RH 2/1442 and RHD 18/394.
88 The extracts came without references, but see almost certainly *M-W*, 25 Jan. 1932, pp. 1014–15, for Fuller's *Grant*, and 11 Apr. 1932, p. 1356, for LH; there are other examples before.
89 See e.g. L. Ritter von Eimannsberger, *Der Kampfwagenkrieg* (Berlin, 1934), p. 110; G.P. von Zezschwitz, *Heigls Taschenbuch der Tanks*, III (Berlin, 1938), passim; Nehring, *Kampfwagen an die Front!*, p. 20 – all identifying the British school with Fuller. In Walter Nehring, *Heere von Morgen. Ein Beitrag zur Frage der Heeresmotorisierung des Auslandes* (3rd edn., Potsdam, 1935), pp. 9, 28–9, LH is presented as Fuller's partner; the conclusion is:

> The actual impact of the inspired enlightening work of these two British officers can be seen in the series of instructive exercises conducted by the British high command in the last four years, which may offer a valuable clue for the armament of a modern army.

90 Fuller's retirement was reported with lavish praise ('the spiritual father of all the ideas of army mechanization of the time') by *M-W*, 4 Feb. 1934, p. 966; 25 May 1934, p. 1525.
91 See Gat, 'Liddell Hart's Theory of Armoured Warfare,' pp. 8–14.
92 LH's article appeared in the *Daily Telegraph*, 3 Oct. 1933; Geyer's covering letter is dated 5 Oct.; both in RH 2/1881. His original report (no. 7) is registered in, but presently missing from, the file in the BA-MA. LH's article was summarized in *M-W*, 18 Dec. 1933, p. 761.
93 23 Apr. 1934, in RH 2/1882.
94 14 Nov. 1934, in RH 2/1882.
95 23 Apr., 2 and 8 May 1935, in RH 2/1883; on 5 Jan. 1935 Geyer reported the appearance of the second edition of Giffard le Q. Martel, *In the Wake of the Tank*.
96 *M-W*, 18 Jan. 1936, pp. 1209–10 (signed 'Captain Schenk'). The important chapter, 'The Future of Armament – and Its Future Use,' was translated in *Wehrgedanken des Auslandes*, 11 (1935), 11:17–26.
97 14 Oct. 1935, 20 pp. plus many appendices, including a sketch of the Tank Brigade's manoeuvre of deep penetration.
98 Abt. III (formerly T3), 'Truppenübungen und Erfahrungen des englischen Heeres 1935,' Nov. 1935, 45 pp. plus maps, in RH 2/1443, quotations from pp. 5, 10; further quotations from LH, pp. 3, 6, 14, 30.
99 Hankey to LH, 27 Dec. 1933, 1/352, probably referring to the British military attaché.
100 For Geyer's correspondence with Berlin, see 15 July 1935, in RH 2/1883; LH's own file of correspondence with Geyer is 9/24/61; invitations were discussed on 10 Aug. 1935, 14 Feb., 16 and 27 May 1936, cited by Bond, *LH*, p. 216, but not by Mearsheimer.
101 Talk with Thorne, 3 June 1942, 11/1942/41; Guderian, *Panzer Leader*, p. 35.
102 Paret to LH, 8 Oct. 1958, in 1/566; he heard it from his father, who reportedly heard it either from Prof. Meinencke or from Sauerbruch.
103 Guderian, *Achtung Panzer!*, p. 167.
104 Nehring, *Kampfwagen an die Front!*, p. 28.
105 Cited from the British cabinet papers by B. Bond and W. Murray, 'The British Armed Forces, 1918–1939,' in A.R. Millen and W. Murray (eds.), *Military Effectiveness*, II (London, 1988), p. 112.

106 Guderian, *Panzer Leader*, p. 32.
107 E. von Manstein, *Aus einem Soldatenleben, 1887–1939* (Bonn: Athenaeum, 1958), pp. 128–9, 240–3.
108 H. Senff, *Die Entwicklung der Panzerwaffe im deutschen Heer zwischen den beiden Weltkriegen* (Frankfurt a.m.: Mittler, 1969), pp. 23–6; N. Reynolds, *Treason Was No Crime: Ludwig Beck, Chief of the German General Staff* (London: Kimber, 1976), pp. 103–5; W. Deist in Militärgeschichtliches Forschungsamt, ed., *Germany and the Second World War*, I (Oxford: Oxford UP, 1990), pp. 431–6; K.J. Müller, *General Ludwig Beck. Studien und Dokumente zur politisch-militärischen Vorstellungswelt und Tätigkeit des Generalstabschefs des deutschen Heeres, 1933–1938* (Boppard a.r.: Boldt, 1980), esp. pp. 208–19; idem., *The Army, Politics and Society in Germany, 1933–1945* (Manchester: Manchester UP, 1987), pp. 5499; Lewis, *Forgotten Legions*, pp. xiii–xiv, 51–3.
109 This is rightly pointed out by Corum, *Seeckt and German Military Reform*, p. 199.
110 Heeresdienstvorschrift 300: *Truppenführung* I (1933), paras. 339–40, pp. 1333–5; II (1934), paras. 725–58, pp. 1–10. See also Reynolds, *Beck*, p. 104, which despite its general sloppiness is the most perceptive about the development of Beck's attitude to armour.
111 Manstein, *Aus einem Soldatenleben*, p. 241.
112 'England: Manöver des Panzerverbandes 18 bis 21.9.1934,' Dec. 1934, copies in RH 2/1442 and RHD 18/394; extracts from the report are printed in Müller, *Beck*, pp. 360–6; see also Reynolds, *Beck*, pp. 104–5.
113 Quoted in ibid., p. 106.
114 Ludwig Beck, 'Nachträgliche Betrachtungen zu dem Einsatz des Panzerkorps in der Lage der Truppenamtsreise vom 13.6.1935,' 25 July 1935: RH 2/v. 134; printed in Müller, *Beck*, pp. 460–5. See also Reynolds, *Beck*, p. 105; Guderian, *Panzer Leader*, p. 32.
115 Kommando der Panzertruppen, 'Erfahrungsbericht über die Versuchsübungen einer Panzerdivision auf dem Truppenübungsplatz Münster im August 1935,' 24 Dec. 1935, RHD 26/2.
116 'Erwägungen über der Erhöhung der Angriffskraft des Heeres,' 30 Dec. 1935, in RH 2/1135; printed in Müller, *Beck*, pp. 469–77; see also Reynolds, *Beck*, p. 105; and RH 2/1224.
117 See e.g. LH's comprehensive article on the future development of the British army in the *English Review*, summarized in *M-W*, 25 Aug. 1935, pp. 315–16, and described as a 'highly interesting study'. The French manoeuvres were covered by a couple of articles in *M-W* in late 1935, and more extensively in 'Überblick über Manöver fremder Heere im Jahre 1935,' *Militärwissenschaftliche Rundschau* [hereafter: *MR*], I (1936), pp. 261–83.
118 The letter by the Allgemeines Heeresamt is dated 22 Jan. 1936, and Beck's reply, 30 Jan., both in RH 2/1135, the latter printed in Müller, *Beck*, pp. 486–90. Again on the three roles of armour, see 23 and 25 Mar. 1936, in RH 2/1135.
119 See the correspondence between Abt. I, II, IV, VIII, and Kdo. der Panzertruppen, in 5 (and no day) May; 3, 15, 17, 18 and 29 June; and (no day) August 1936: all in RH 2/1135; 2 May 1936, in RH 2/1382.
120 Guderian, 'Schnelle Truppen einst und jetzt,' *M-W*, IV (1939), pp. 240–1. See also Guderian, *Panzer Leader*, p. 36; Manstein, *Aus einem Soldatenleben*, p. 242.
121 Müller, *Beck*, esp. p. 474; idem., *The Army, Politics and Society*, p. 85.
122 See Abt. II, 13 Aug. 1936, in RH 2/1135, and cf. Guderian, 'Die Panzertruppen und ihr Zusammenwirken mit den anderen Waffen,' *MR*, I (1936), pp. 614–15 (citing Fuller, *The Army in My Time*); reissued separately in book form the following year.
123 Beck had already raised the issue in his 30 Dec. 1935 memorandum, in RH 2/1135, printed in Müller, *Beck*, p. 472. For his decision in the affirmative, see Abt. II, 5 and 12 Oct. 1936, in RH 2/1135.

124 Manstein, *Aus einem Soldatenleben*, p. 243ff.
125 Guderian criticized the change in British policy, which must have been waved in his face in Germany: 'Die Panzertruppen und ihr Zusammenwirken mit den anderen Waffen,' *MR*, I (1936), pp. 614–15.
126 This distinction was first developed in regard to the British army by H. Winton, *To Change an Army* (Lawrence: Kansas UP, 1988).
127 For comprehensive surveys, see e.g. *M-W*, 11 July 1932 (signed 'Lieutenant Freytag'); Nehring, *Kampfwagen an die Front!*, pp. 25–6; also Nehring, *Heere von Morgen*, pp. 38–40, citing the Soviet theorist Isserson (Moscow, 1932); Walter Nehring, *Panzerabwehr* (2nd edn., Berlin: Mittler, 1937), also citing Soviet sources and views, including those of Tukhachevskii; *Heigls Taschenbuch der Tanks* (2 vols., Munich: Lehmann, 1935), pp. ii, 455–7; M.J. Kurtzinski (ed. and trans.), *Taktik schneller Verbände* (Potsdam, 1935); internal staff reports: 14 Mar. 1935, in RH 2/1438; 1 Apr. 1936, in RH 2/1439; 14 Sept. 1936 and 8 Apr. 1937, in RH 2/1444; *M-W*, 11 Oct. 1936, pp. 720–1 (from *The Times*); 18 Oct. 1936, pp. 776–7; 11 Dec. 1936, pp. 1188–93; 25 Dec. 1936, pp. 1332–5; Jan. 1937, pp. 1589–92; *MR*, III (1938), pp. 557–74, 671–86.
128 Guderian to LH, 24 Jan. 1949, in 9/24/62, mistakenly dating the visit to 1933; Nehring, *Panzerwaffe*, pp. 42–6; Zeidler, *Reichswehr und Rote Armee*, p. 196.
129 Heinz Guderian, 'Kraftfahrkampftruppen,' *MR*, I (1936), p. 73, citing Kryshanowski; also Guderian, *Achtung Panzer!*, pp. 151–4.
130 Nehring, *Kampfwagen an die Front!*, p. 26; Guderian, *Achtung Panzer!*, p. 153; Heinz Guderian, 'Schnelle Truppen einst und jetzt,' *MR*, IV (1939), pp. 237–8. The German criticism of the Soviet armour doctrine in the 1930s has been well noted by Senff, *Die Entwicklung der Panzerwaffe*, p. 22.
131 For the radio, see Macksey, *Guderian*, pp. 50–1, 67.
132 Guderian, *Panzer Leader*, pp. 38, 46; Nehring, *Panzerwaffe*, p. 93; the material relating to the manoeuvres can be found in RHD 18/296–8, RHD 18/357–8; the detailed report of Major-General A. C. Temperley on the manoeuvre and the state of the Panzer arm for the *Daily Telegraph* and *Morning Post* was translated into German and sent to Berlin by the German military attaché in London: RH 53–7/v. 54.
133 See esp. F.O. Miksche, *Blitzkrieg* (London: Faber & Faber, 1942); Klaus Maier's contribution to Militärgeschichtliches Forschungsamt, ed., *Germany and the Second World War*, II (Oxford: Oxford UP, 1991), pp. 41–3.
134 RH 2/2930.
135 RH 2/1437–41.
136 *M-W*, 25 Nov. and 4 Dec. 1936, pp. 1059–63, 1108–10; for the armour, see pp. 1109–10.
137 *M-W*, 22 Oct. 1937, p. 1053; also 29 Oct., pp. 1117–18.
138 Zina Hugo to LH, 28 Oct. 1941, 13/5 miscellaneous file, also 11/1948/38; cited by Bond, *LH*, p. 230, but not by Mearsheimer. See also LH, *The Other Side of the Hill* (London: Cassell, 1951), pp. 65–6.
139 The Thoma file, 1 Nov. 1945, 9/24/144; Guderian to General Dittmar, 29 Aug. 1948; Guderian to LH, 7 Oct. 1948 and 19 Mar. 1949: all in 9/24/62; none cited by Mearsheimer, but see Bond, *LH*, pp. 229–31.
140 Guderian to LH, 19 Mar. 1949, 9/24/62.
141 Mearsheimer, *LH*, pp. 33–46; but see Gat, 'Liddell Hart's Theory of Armoured Warfare.'
142 LH to Chester Wilmot, 14 May 1953, 9/24/30; cited by Mearsheimer, *LH*, 161.
143 Macksey, *Guderian*, p. 41; idem., *The Tank Pioneers*, p. 118; Mearsheimer, *LH*, pp. 165–6. Guderian, *Achtung Panzer!*, p. 141, cites Fuller, Martel, and Liddell Hart, in that order, as the pioneers of the idea of an all-arms armoured formation.
144 Macksey, *The Tank Pioneers*, p. 118.
145 Fuller, *Memoirs*; Swinton, *Eyewitness*; Martel, *In the Wake of the Tank*.
146 Guderian, *Achtung Panzer!*, pp. 141, 170.

147 Macksey, *Guderian*, p. 69.
148 For this, see Gat, 'Liddell Hart's Theory of Armoured Warfare.'
149 *Army Quarterly* (Aug. 1942), p. 216, cited in J.F.C. Fuller, *Armoured Warfare* (London: Eyre & Spottiswoode, 1943), p. 5.
150 A.J. Trythall, *'Boney' Fuller* (London: Cassell, 1977), pp. 184, 187, 192, 203.
151 Quoted in Macksey, *Guderian*, p. 41. For the Fuller-Guderian meeting, see also Trythall, *Fuller*, p. 203.
152 Guderian's son was himself a junior officer in the Panzer force in the late 1930s, but he was not his father; there are the problems of the distance of time and of the inquirer's leading questions, and there is the weight of other evidence of the time.
153 See e.g. *M-W*, 20 Apr. 1936, pp. 1756–9 (signed '326'); 11 June 1937, pp. 3066–7; 29 Oct. 1937, pp. 1117–18 (quotation; signed '326'); for the *Times* originals see Gat, 'Liddell Hart's Theory of Armoured Warfare,' pp. 11–12. See also the intelligence translation: Abt. III, 12 Apr. 1937, RH 2/1440.
154 L. Addington, *The Blitzkrieg Era and the German General Staff, 1865–1941* (New Brunswick, NJ: Rutgers UP, 1971); M. Geyer, *Aufrüstung oder Sicherheit. Die Reichswehr in der Krise der Machtpolitik, 1924–1936* (Wiesbaden: Steiner, 1980), p. 481; M. Carver, *The Apostles of Mobility* (London: Weidenfeld, 1979), pp. 63–4; D.J. Hughes, 'Abuses of German Military History,' *Military Review*, 66 (Dec. 1986), pp. 69–70; also, somewhat along these lines, Mearsheimer, *LH*, pp. 87, 92. For the argument that Blitzkrieg diverged from the traditional German emphasis on battles of annihilation, see M. Cooper, *The German Army, 1933–1945* (London: Macdonald & Jane's, 1978), pp. 130–48 and passim; J. Mearsheimer, *Conventional Deterrence* (London: Cornell UP, 1983), pp. 38–9; B. Posen, *The Sources of Military Doctrine: France, Britain, and Germany between the World Wars* (London: Cornell UP, 1984), pp. 86, 206–7.
155 Guderian, *Panzer Leader*, p. 316.
156 Michael Geyer's hyperbolic claim that Blitzkrieg was defined as an operational design only in hindsight and with some help from Liddell Hart is basically correct: Geyer, 'German Strategy,' pp. 585–6; also M. Messerschmidt, 'The Political and Strategic Significance of Advances in Armament Technology: Developments in Germany and the "Strategy of Blitzkrieg",' in R. Ahmann, A.M. Kirke, and M. Howard (eds.), *The Quest for Stability* (Oxford: Oxford UP, 1993), pp. 249–61; J.P. Harris, 'The Myth of Blitzkrieg,' *War in History*, 2 (1995), pp. 335–52, repeating, however, the erroneous claims about LH's lack of influence.
157 *M-W*, July 1940, p. 165; Guderian, *Panzer Leader*, p. 461. For LH, see *The Defence of Britain*, p. 101.
158 Robert O'Neill, *The German Army and the Nazi Party* (London: Cassell, 1966), p. 127.
159 Quoted without reference by Macksey, *Guderian*, p. 59.
160 Nehring, *Heere von Morgen*, p. 31.
161 Guderian, *Panzer Leader*, esp. pp. 92, 159, 166–9, 182, 185, 199; Guderian to LH, 24 Jan. 1949, 9/24/62.
162 Germains had anticipated this criticism in his *The 'Mechanization' of War*, p. 178. For North Africa, see the opinions of Rommel and Bayerlein, *The Rommel Papers*, pp. 159, 184. M. Carver, *Tobruk* (London: Dufour, 1964), esp. pp. 254–5, lays much of the blame on LH's teachings; also S. Bidwell, *Gunners at War* (London: Arms & Armour, 1970), pp. 163–82. In his reply to Carver LH cleverly threw the blame on the 'cavalry mentality' of many of the newly converted British armour commanders: *Times Literary Supplement*, 19 Nov. 1964.
163 *M-W*, 6 Feb. 1942, p. 918; for the original, see LH, *This Expanding War* (London, 1942), p. 171 (18 Jan. 1942).
164 For Beck, see again Dec. 1934 in RH 2/1442, cited in Müller, *Beck*, p. 361; also e.g. Nehring, *Kampfwagen an die Front!*, pp. 18–19.
165 Heinz Guderian, 'Schnelle Truppen einst und jetzt,' *MR*, IV (1939), p. 243.

166 Cited without reference in Ogorkiewicz, *Armoured Forces*, p. 21.
167 See e.g. Nehring, *Heere von Morgen*, bib., pp. 9, 33–5; idem., *Panzerabwehr*, p. 53; Heinz Guderian, 'Kraftfahrkampftruppen,' *MR*, I (1936), p. 71; idem., 'Die Panzer-truppen und ihr Zusammenwirken mit den anderen Waffen,' *MR*, I, pp. 616–17, 619; Guderian, *Achtung Panzer!*, p. 150, bib.
168 Thoma to LH, 1 Nov. 1945, 9/24/144; Guderian to LH, 7 Oct. 1948, 9/24/62; Nehring, *Panzerwaffe*, p. 85.
169 Quoted in Trythall, *Fuller*, pp. 209–10.
170 L. Ritter von Eimannsberger, *Der Kampfwagenkrieg* (Munich: Lehmann, 1934), pp. 109–10, 167.
171 Ibid., pp. 113–209.
172 Ibid., pp. 170–209.
173 ibid., pp. 161–9.
174 D. Bradley, *Generaloberst Heinz Guderian und die Entstehungsgeschichte des modernen Blitzkrieges* (Osnabrück: Biblio, 1978), pp. ii, 184–7, solicited the evidence of some of the Panzer arm's veterans. The verdict was shared by all of them, including Nehring, whose works in the 1930s regularly referred to Eimannsberger among the other authorities cited: Nehring, *Heere von Morgen*, source list and pp. 43–4; idem., *Panzerabwehr*, p. 13.
175 In the second edition of his book (1938) Eimannsberger introduced a number of other changes to conform to the divisional structures actually adopted by the Wehrmacht. Miksche, *Blitzkrieg*, pp. 107–9, relying on this edition, mistakenly believed that these were Eimannsberger's original proposals of 1934, thus forming an exaggerated idea of his potential influence.
176 For material relating to the first manoeuvres of the Panzer divisions, see RHD 26/25.
177 Ibid.; Heinz Guderian, 'Kraftfahrkampftruppen,' *MR*, I (1936), p. 68; idem., *Achtung Panzer!*, p. 143; the whole thing was rightly pointed out by Ogorkiewicz, *Armoured Forces*, pp. 43–4, 578–9.
178 In addition to the reports mentioned above, see e.g. Nehring, *Heere von Morgen*, pp. 32–3, 72–3; Guderian, 'Kraftfahrkampfwagen,' pp. 68–9; again, see Ogorkiewicz's perceptive comments in *Armoured Forces*, pp. 72–3.
179 For British opinions here, see Gat, 'Liddell Hart's Theory of Armoured Warfare,' pp. 17–22; Guderian, *Achtung Panzer!*, p. 141.
180 Quoted in Macksey, *Guderian*, p. 59.
181 Summary of *Paris* in *KGA* II (1926), 8:23–6; 'The Remaking of Modern Armies,' *Daily Telegraph*, summarized in *KGA* III (1927), pp. 18–23.
182 A review article of the book ('The Well-Known Captain Liddell Hart'): 'novel surprising proposals, some of which are already implemented in experimental units': *M-W*, 11 Jan. 1934, pp. 859–61; also *M-W*, 4 Oct. 1934, p. 513. For the book's resonance in Germany, see *M-W*, 18 Jan. 1936, p. 1209; Nehring, *Heere von Morgen*, bib., p. 15. An interesting example is the book's influence on Felix Steiner, a battalion and regiment commander in the Waffen SS and developer of training methods for infantry: B. Wegner, *The Waffen SS: Organization, Ideology and Function* (Oxford: Blackwell, 1990), p. 184. That LH's book had in fact little to say on armoured infantry has been noted by B.H. Reid, *J.F.C. Fuller: Military Thinker* (New York: St. Martin's Press, 1987), p. 161.
183 See W. Velten, *Das deutsche Reichsheer und die Grundlagen seiner Truppenführung* (Bergkamen: AGEMA, 1994); and the persistent emphasis in the training instructions issued by the Command of the Panzer Troops: RHD 26/2 (24 Dec. 1935), RHD 26/3 (10 Nov. 1935), RHD 26/4 (10 Nov. 1936), RHD 26/5 (15 Nov. 1937); also Guderian, *Achtung Panzer!*, p. 178ff. and passim.
184 For LH's views on the subject, see e.g. *M-W*, 25 Aug. 1935, pp. 315–16; and of course in the translated *When Britain Goes to War*.

185 Guderian, *Panzer Leader*, p. 37.
186 The expected reduction in the tank establishment of the British armoured division from its earlier very high level was noted by Guderian on the eve of the war: 'Schnelle Truppen einst und jetzt,' *MR*, IV (1939), p. 238.
187 For LH, see Gat, 'Liddell Hart's Theory of Armoured Warfare,' pp. 22–3.
188 See e.g. F. von Senger and Etterlin, *Neither Fear Nor Hope* (London: Macdonald, 1960; introd. by LH), pp. 79–80.
189 Guderian, *Panzer Leader*, pp. 294–5, quoting his conference notes of the time; he probably referred to LH's article of 31 Dec. 1942: Gat, 'Liddell Hart's Theory of Armoured Warfare,' pp. 23–4. The authenticity of Guderian's evidence is beyond question, for he had not been advised on this point by LH, who was himself surprised and delighted to read it in Guderian's memoirs: LH to Broad, 28 Dec. 1951, 1/108.
190 26 May and 15 June 1942, *The Rommel Papers*, p. 203.

2 Technology, national policy, ideology, and strategic doctrine between the world wars

In the wake of the Second World War the controversies of the interwar period regarding both national policy and strategic doctrine were dramatically viewed as struggles between prescience and folly. Winston Churchill and Neville Chamberlain were the pair that epitomized the political debate in the West. J.F.C. Fuller, B.H. Liddell Hart, Hans Guderian, Charles de Gaulle, and Billy Mitchell were some of the names that represented the struggle against 'conservative' military establishments. The 'heroes' dominated public and historical consciousness, indeed, wrote much of the historiography of the period themselves. Their narratives continue to dominate the popular view and media. However, from the 1970s, as these towering figures were passing away and national archives were opening, scholars have been crystallizing a more complex and nuanced picture, in which 'right' and 'wrong' have not been as starkly contrasted as before. There can be no dispute that the decisions with which the people of the interwar period were grappling were indeed dramatic and fateful, and not all options proved *equally* realistic in terms of their proponents' aims and existing capabilities. Still, these options represented real dilemmas, deep constraints, genuine uncertainties, and internally conflicting goals, which being famously the stuff of life in general, seem to have been all the more acute in the interwar period, particularly the 1930s. This article will start with the problems of technology and strategic doctrine, then proceed to examine how these were to be squared with particular national aims, power, and geo-strategic conditions.

The technological horizon

In outline and abstract the nature of the overall technological transformation that challenged the armed services during the interwar period was the most general constant underlying the formulation of strategic doctrine. Modernist theorists and military men who enthusiastically embraced the vision of a coming industrialized and mechanized new age were particularly receptive to the change. During the first decades of the twentieth century many of these men belonged to the futuristic-technocratic-technological strands of proto-fascism and fascism (Giulio Douhet, Fuller, Ernst Juenger), conscientiously allied with fascism (Werner Blomberg, Walter Reichenau, Heinz Guderian), were sympathetic to it (Henry Ford), or at any rate felt that it was the wave of the future (Charles Lindberg).[1] However, even the proverbial 'conservative' officers, steeped in the traditional agrarian and gentlemanly world, were on the whole grudgingly recognizing the changing reality, even if they could not define it in terms of the general historical transformation of the age, as Fuller, for example, did.

Fuller saw that the process affecting the armed services was the same one that had been transforming society as a whole, namely, the unfolding of the Second Industrial Revolution that had been gathering momentum from the end of the nineteenth century. Earlier during the nineteenth century it had been the principal products of the First Industrial Revolution – the steam engine, machine tools, and advanced metallurgy – that revolutionized navies and armies, as all other spheres of life. Iron and steel steam ships and steel, rifled breech-loading guns had revolutionized naval warfare. On land, the same steam engine, as applied to railroads, had revolutionized strategic mobility and logistics, while steel, rifled, breech-loading, and magazine-fed 'repeaters', rapid-firing guns, and machine guns had revolutionized tactics. The Second Industrial Revolution involved a new cluster of technologies, most notably the internal combustion engine, chemicals, and electricity. Like the First Industrial Revolution it was to take decades to unfold, transforming human life in general, including the armed forces. In the span of one decade (1895–1905), various applications of the internal combustion engine inaugurated forms of mobility that had not been possible with the steam engine: the automobile on land, the airplane in the air, and the submarine at sea. All had their military versions before the First World War, and they rapidly grew to dominance by its end. Similarly, the chemical industries produced the poison gases that clouded the battlefields of the Great War. Electricity, in the shape of electric telegraphy, had already revolutionized naval and military communications – as communications in general – during the second half of the nineteenth century. The more advanced system, radio communication, was entering the naval and military fields before and during the war.

As Fuller saw it, the future was on track. All the innovations of the Second Industrial Revolution as applied to warfare and prominently revealed during the Great War had not yet run their course. The technologies had not yet fully matured and their absorption process had been far from being completed. Predicting the future was almost a matter of simple extrapolation from ongoing trends – in the sense that complex things are simple for geniuses like Fuller. By 1923, in his *Reformation of War*, the whole picture had been clear to him. On land, the light and flexible internal combustion engine would conquer the countryside, as it had been doing in civilian life in the car, the truck, and the tractor. Mechanized power and mechanized mobility, which railroads had revolutionized for strategic deployment and logistics alone, would now revolutionize operations and tactics on the battlefield itself. Muscle power was out. All the traditional arms would transform to join tanks on tracked armoured vehicles or disappear altogether, except on special ground. In the air, aircraft would cross countries, bombing the enemy's heartland. At sea as well, the applications of the internal combustion engine in the shape of the aircraft and submarine, together with flotilla boats, would dominate, driving the mammoth battleships to extinction. Poison gas would be extensively used in all theatres of war. Communication with the mechanized fleets would be kept by radio, more advanced than the heavy and barely mobile equipment of the First World War. Furthermore, by 1928 Fuller was already envisioning electric-automated-robotic warfare of the future, whose full application, as we know, was only to unfold with the coming of the third of the great waves of industrial technology, that of electronics, which we are still experiencing today in the military as well as in the civilian spheres.[2] Many other people shared at least parts of this vision, but probably none formulated it so comprehensively and contextualized it historically in such an illuminating manner as Fuller did. As Guderian was to write about Fuller and his disciples: 'They envisaged it [the tank] in relation to the growing motorization of our age, and thus they became the pioneers of a new type of warfare.'[3]

However, as mentioned earlier, all this was in outline and in abstract. In outline, because the prophets of mechanization offered sweeping visions that in large part proved amazingly prescient; still, some of their predictions proved less happy, some conflicted with one another, and there was, finally, a great deal of detail to be worked out in the immensely complex and laborious process of turning vision into reality. By the 1930s, as rearmament was taking place among all the great powers (despite the radicals' indignant rhetoric, there had been no perceived threat and no money before that), few if any reactionary officers were prominent anywhere. Harold Winton's classification of the positions within the British army with respect to the mechanization of land warfare was true of all the armed services in all countries.[4] There were groups of active radicals on one end and mostly inert reactionaries on the other, but the majority of officers who had an opinion probably belonged to a broad centrist spectrum of progressives to mildly conservative. These were very conscious of the transformation of warfare by new technologies but were not sure as to how radical this transformation would be and the form it would take. The weight of vested organizational interests and bureaucratic inertia ought never to be underestimated, but it was not the main factor.

Thus, on land, practically nobody disputed that the tank and mechanized troops would play a decisive role in future war. But how 'almighty' would they be? First, would not the proliferation of anti-tank guns and mines do to the tank what the magazine rifle and machine gun had done to the cavalry and infantry? Even the prophets of armoured forces like Fuller (as early as 1928), Liddell Hart, Mikhael Tukhachevsky, and Guderian (all by the mid-1930s) recognized that a new balance between tank and anti-tank was forming. While they insisted that this made the concentration of the armoured forces to overwhelm the enemy's dispersed and immobile defences all the more necessary, some of them realized that the process might again lead to stalemate and attrition warfare. The Blitzkrieg campaigns of 1939–1941 should not mislead; for this is precisely what happened in the later part of the Second World War, from 1942 on, once all armies acquired tanks and anti-tank weapons and learnt how to use them. Largely mechanized but semi-mobile, materiel-attrition warfare would now prevail rather than either the unmechanized and immobile mode of the early World War I or the elite-mechanized, mobile, and decisive type projected in the 1920s and realized in 1939–1941.

Further, while all wanted tanks in the 1930s, and in large numbers, the dividing line between 'radicals' and 'progressives', on the one hand, and 'conservatives', on the other, was on whether specialized armoured formations should be created for a concentrated mobile armoured punch. Here Britain was the world leader in both theory and practice from the mid-1920s, and her example was eagerly studied and variably imitated by all the great powers.[5] Eventually, even the French created (four) armoured divisions (*Division Cuirassée de Réserve*) on the eve of World War II. But then, the dividing line between the 'radicals' and 'progressives' was over the *exclusiveness* of the armoured formation, and, indeed, over the future of the *non*-mechanized forces, at least in the near, practical, future, for which armies, as opposed to visionary theorists, must prepare. For reasons that will be discussed more fully later on, the British 'radicals' saw little if any future for non-mechanized forces in a great power war, even in the short run. The leaders of the Panzer arm, mostly middle-rank officers, were preoccupied with their creation and had little time for the rest of the army. Only in the Soviet Union did the proponents of Deep Battle – who were senior enough themselves to occupy the army's high command – formulate their ideas as an official doctrine from above for the entire army, combining mechanized and non-mechanized elements. All the same, in all of the powers non-mechanized forces remained the bulk of the armies.

Limited industrial capacity above all else made a fully mechanized army a very distant objective in the 1930s, even if army high commands had regarded it as a desirable objective, which was not at all clear. 'Progressives' accepted the need for independent armoured formations but took issue with the 'radicals' on whether tanks and mechanized vehicles in general should be solely directed to armoured divisions or should also be distributed to support non-mechanized infantry, as well as used in 'light divisions' for cavalry-type missions like reconnaissance, screening, and pursuit. In the absence of combat experience other than the First World War, there was no way of knowing in advance what would prove more effective, and no responsible army authorities could or did put all their eggs in one basket. By the mid-1930s, all the great powers' armies opted for both armoured divisions and tanks for infantry support, all motorized some of their infantry divisions on trucks, and some created light mechanized divisions as well.

Even sensible ideas and prudent middle-of-the-road solutions can have very unfortunate consequences. The main consequence concerned the diversification of tank production between models intended for the armoured formations and the slow and heavily tanks for close support of unmechanized infantry, models which lacked the range and speed for effective participation in mobile operations. Once the armies possessed both tank types and, more importantly, once industry was geared to produce them, any retreat from this policy was difficult to carry out. The British, who had pioneered the concept of the general, 'medium' battle tank in the 1920s, adopted the 'cruiser' and 'infantry' types in the mid-1930s, a schism that plagued their armoured operations throughout World War II. The French as well, in 1939–1940, found that the heavy infantry models could not be converted to mobile warfare, even when hastily assembled into independent formations after the demonstrated success of the Panzer divisions in Poland. In theory, the Red Army had so many tanks – as many the rest of the world put together – to be able to afford its elaborate division in tank design between close-range infantry support, medium-range infantry support, and independent long-range missions. In practice, it was only able to survive this division because the Soviet Union was so much larger and stronger than France; but practically its entire tank fleet was destroyed in the summer of 1941. Deep Battle was the most advanced and comprehensive operational doctrine developed in the 1930s; but as the Red Army revived it later in the war and continued to use it until the disintegration of the Soviet Union, the concept would now centre on the main battle tank.

The Germans were scarcely wiser than the others. The progressivist Chief of the General Staff Ludwig Beck, who initiated the creation of the first three Panzer divisions in 1935, wanted even more tanks in both infantry support tank brigades and light mechanized divisions. He also wanted the design and production of special infantry tanks, a plan that failed to materialize because Germany's production capacity was already strained to support the massive rearmament of its armed services.[6] Paradoxically, this proved to be a crucial factor behind the German advantage in armoured warfare at the outbreak of the Second World War. When the Polish campaign demonstrated that the most effective use for the tanks was in Panzer divisions, the Wehrmacht before the Battle of France experienced no trouble in converting its light divisions and infantry support tank brigades, which had been using the same tank types as the Panzer divisions.

Finally, the doctrine and composition of the independent armoured formations themselves involved many open questions, not for all of which the prophets of armoured warfare had magic answers, or the right ones. How would the envisaged mechanized fleets operating deep behind the enemy's front be controlled and supplied? Radio was the answer to the first question, but the British only developed and distributed the necessary light sets in the late 1920s to early 1930s, and the Germans in the second half of the 1930s, while the French and Soviets lagged far behind. As for supplies, the British theorists, following in Fuller's footsteps, believed unrealistically that the mechanized formations should and could be logistically self-sufficient. They would have no lines of communications, carry everything they needed with them, as well as live off the countryside using enemy civilian gas stations, and in emergency have critical supplies flown to them.

Turning to the most famous problem, everybody agreed (again following in Fuller's footsteps) that the armoured division ought to include all arms, but in what ratio and for what functions? Fuller and most of the British armour pioneers held that infantry would only have a defensive, second line, role to play in the armoured division. Liddell Hart, diverging from Fuller on this point, insisted on some offensive role for the mechanized infantry in close cooperation with the tank spearheads. He wrongly maintained, however, that a very small component of infantry would suffice for the armoured division. Furthermore, like Fuller he insisted that only fully tracked and armoured, rather than trucked or bused, infantry was to be incorporated – as should be the norm for the other arms as well – because all the components of the armoured division ought to have equal cross-country mobility. This was indeed desirable in principle, except that no country in the 1930s could afford to mount all the elements of the armoured division on armoured cross-country vehicles, and all had to use wheeled transport instead. The British armour pioneers never accepted this, and during the late 1930s their disappointment only reinforced their tendency to advocate very tank-heavy armoured divisions, in which all the other elements were much reduced. Guderian and the German pioneers resigned themselves to the facts of life when building the Panzer divisions on the model of the balanced 1934 British experimental Mobile Division. Only the authors of Deep Battle consciously decided from the start that the Red Army would for a long time have to combine tracked, wheeled, and non-motorized elements.[7]

In the air, the situation was similar, even though some of the winning ideas were instructively opposite to those that proved successful in land warfare. As on land, for all the frustration of the air power pioneers, the lack of any serious perceived threat during the 1920s meant that there was no money for the armed services and very little investment in the air. However, as rearmament began in the 1930s, the recognition that air power would be central, indeed paramount, in future war was universal, dominating the media, public opinion, governments, and general staffs. In fact, it bordered on a general panic, for assessments of the efficacy of air bombing were vastly exaggerated, even when the later and secret invention of radar, with the enormous boost it would give the defence, is factored in. The threat was viewed in terms that recall the fear of a general holocaust in the nuclear age.

Indeed, the main question to be settled was how dominant air power was and what form exactly it should take. The most radical prophets, like the Italian Giulio Douhet before, during, and after the First World War (and H.G. Wells even before him), believed that air power was so effective that it was able to bomb into destruction the infrastructure of a great power within hours or days. It therefore rendered the other services – the army and navy – superfluous. The American William (Billy) Mitchell, under Douhet's influence, was similarly radicalizing in the second half of the 1920s. By the end of World War I and again in the 1930s, the leaders of the RAF, the world's first independent air force, were only scarcely less radical. Indeed, air forces tended to stress their 'independent' role as their *raison d'être*, and generally resisted the allocation of aircraft to any other mission or armed service. Douhet envisaged a standard heavy bomber and no other aircraft type.

All the same, as on land, the responsible authorities – both civil and military – though highly impressed by the threat of air power, were again not going to gamble everything on an untried and unproven card. Indeed, their diversified, compromise building programmes proved more realistic, a result that was much in contrast to the 'radical'/'progressive' tank types debate on land. There is no easy lesson here. Thus, while giving preference to 'strategic' bombing, practically all the great powers designed heavy and medium bombers, fighters, and aircraft for close support in the land battle and for participation in naval warfare. As we shall see, it was particular national circumstances that were responsible for the changes of emphasis that developed between the great powers' construction programmes from the mid-1930s.

The 'terrible ifs', as Churchill called them, were hanging over sea warfare as well. In the wake of World War I respected admirals such as William Fullam, William E. Sims, and Bradley Fiske in the US Navy and known radicals like Herbert Richmond in the Royal Navy held the views already cited here from Fuller: aircraft, aircraft carriers and submarines, together with flotilla boats, would dominate future warfare at sea, leading to the demise of the battleship. By the late 1920s Billy Mitchell moved farther from this position, writing that land-based aircraft would be enough to destroy all surface ships, making even aircraft carriers unnecessary.[8] Still, as navies were building up again in the 1930s to face the renewed threat of war, and after discarding the limitations of the Washington and London treaties, *all* of them adopted balanced construction programmes. Who would dare to run the potentially catastrophic national risk of falling behind in the race for the construction of battleships, which had so long been the measure of naval power, let alone recklessly discard them?

Indeed, did battleships actually become wholly outdated before the introduction of radar in the very late 1930s and early 1940s made it possible for navies to avoid close-range, unexpected encounters, even in bad weather and at night? Thus, the major navies built aircraft carriers in large numbers, while at the same time giving top priority to battleship construction. If the British navy's air arm was behind those of other leading navies, it was largely because until 1937 it had belonged to the RAF, for which it had been an even lower priority. Top priority to the capital ships in resource allocation also came dangerously at the expense of the anti-submarine flotilla boats, with whose vital necessity the Royal Navy was all too familiar from the previous war. But then, as naval officers demonstrated to a hopeful Churchill, did not the newly invented sonar (ASDAC) spell the decline of the submarine?[9] Only experience in large-scale warfare could decide all these questions. Since such a test was very infrequent, the armed services had to make their choices for the future in a reality that was even more uncertain than it was in other spheres of life. It is largely in hindsight that some decisions rather than others regarding doctrine and construction appear as blunders in our eyes.

National aims, means, and geopolitical conditions

The sweeping technology-based visions of future warfare did not only leave open a multitude of major and minor details to be filled in on their bare outline; they also had to be applied from the abstract to take account of the particular aims, means, and geopolitical conditions of the various powers. The strategic challenges facing each of the great powers in the 1930s were no less than existential (with the exception of the United States, where they were also very serious), but they were differently modulated.

The acute tensions inherent in the German case have received considerable scholarly attention.[10] Germany was the single most powerful state in Europe, but at the same time it was a middle-size country, with a limited resource base.

Furthermore, the First World War taught it that any major attempt on its part to revise the status quo by force was likely to create a superior great power coalition against it. Worse still, its defeat meant that its status as a great power had been strictly limited by the victors. A wide consensus in Germany sought the revision of the Versailles Diktat, by peaceful or more aggressive means. Hitler's rule meant the more ambitious goal of European hegemony and world power. What strategic doctrine could support either the limited or unlimited policies of revision? How was the tension between great power, relative overall inferiority, and revisionism to be squared? One of the main lessons that Germans drew from the First World War was that every effort had to be made to harness all available means – material, social, and moral – to the war effort, so that the most is made of Germany's inferior resources in the age of total war. Total mobilization and a cohesive *Volksgemein-scahft* (popular national community) were deemed essential. From Ludendorff to Hitler these were associated with a radical regime that would be able to galvanize and regimentalize the masses. For that reason, some high-ranking officers, such as Blomberg and Reichenau, and many of the younger, middle-rank ones of the generation of Guderian and the war-hero-novelist-military-theorist Ernst Juenger, broke with the officer corps' famous class conservatism and looked for radical political-social solutions to radical political-strategic dilemmas.[11]

Indeed, total mobilization and *Volksgemeinschaft* were regarded as essential but insufficient to cope with Germany's political-strategic dilemmas. Its rivals as well were expected to mobilize all their resources and again, as in the First World War, impose on Germany a protracted attrition conflict and economic blockade that would gradually bleed it white. To prevail, Germany would have to avoid attrition conflict by devising ways of rapidly overwhelming its enemies. Here too, it is no coincidence that political and strategic radicalism went hand in hand. For a very long time after the Second World War a myth prevailed that in the 1930s Germany's leaders had conceived the so-called Blitzkrieg doctrine, which both militarily and economically had relied on rapid, short engagements rather than on a protracted total war. In reality, Richard Overy and others have shown that economically Hitler and the Nazi regime prepared for a protracted struggle and total mobilization.[12] In the military sphere as well – from which the term Blitzkrieg was derived – it has slowly dawned on scholars some half a century after the war that nothing like a Blitzkrieg doctrine existed in the German army before the war, either officially or unofficially. Although the words are German for 'lightning war', the term was of foreign origin, invented by the media at the opening stage of the Second World War.[13]

All the same, while no Blitzkrieg doctrine existed either in the German army or economy, notions regarding the need for swift and unconventional coups were inevitably present in view of the nature of Germany's strategic dilemmas. Radicals like Blomberg, chief of the army's covert general staff (*Truppenamt*) in the late 1920s and later war minister under Hitler, were most notable in looking for radical operational schemes and revolutionary doctrines, which *inter alia* made them extremely interested in the possibilities opened up by mechanization. As early as 28 February 1934, Hitler, addressing army and Sturmabteilung leaders at the

Reichswehr ministry, predicted that in order to gain living space in the East for the German people in the teeth of international opposition, 'short, decisive blows to the West and then to the East could be necessary'.[14] It is no wonder that when Guderian famously gave him a demonstration of the mechanized troops at the army's demonstration ground for weapon development at Kummersdorf, Hitler repeatedly uttered: 'That's what I need! That's what I want to have!'[15] Although Hitler's involvement in the army's affairs was very limited before the war, the mechanized troops were known to have been held in special esteem by the regime and by the radical war ministry (Blomberg and Reichenau), something which at least to some degree helped their cause in the army. As the mechanized troops demonstrated their potency in Poland and, later, France, Blitzkrieg was born, still not an official or written doctrine but nonetheless the German standard concept and practice.

Much the same applied to the development of the Luftwaffe, which would become the Panzer forces' partner in executing Blitzkrieg. As an independent service headed by Herman Goering, Hitler's second in the Nazi hierarchy, the newly established Luftwaffe, like all air forces, regarded independent strategic bombing as its principal mission. Its first chief of staff Walter Wever wanted heavy, four-engine bombers. This building programme did not materialize due to engine production problems; because Hitler and Goering opted for rapid acquisition of a larger number of medium, twin-engine bombers, for both military and propaganda/deterrence/blackmail aims; and because the leaders of the Luftwaffe reckoned that in wars in Central and Western Europe medium bombers would have sufficient range for strategic bombing as well.[16] After the war, the German decision to delay on the heavy bomber was criticized by commentators, a criticism more recently resumed by Richard Overy. It has been argued that this decision crippled the Luftwaffe in the Battle of Britain and the Battle of the Atlantic, and made the bombing of the huge Soviet production facilities in the Urals impossible. However, with the exception of the Battle of the Atlantic, where a small number of long-range aircraft was critical for locating the Allied convoys for the German U-boats (the FW 200 came too late and were too few), it is difficult to make sense of this argument. 'Strategic' bombing was notoriously a very dubious instrument of war in terms of cost/effectiveness for the Western Allies as well. Its main strategic and political justification was that until 1943, or, indeed, the summer of 1944, the Allies had no major land theatre to engage the Germans and take the pressure off the Soviets. Furthermore, whatever fruits the bombing campaign brought the Allies after years of sustained effort, they were only achieved by a gigantic construction programme based on the resources of the world's mightiest industrial power by far, the United States.

None of these conditions applied to Germany. By contrast to the Western Allies, Germany was first and foremost a continental power, whose most cost-effective strategy was to destroy its enemies' armed forces and occupy their territory. Although, as mentioned earlier, the Luftwaffe's leaders envisaged it as an independent, 'strategic' force, a claim that was never given up, the Second World War (and the Spanish Civil War before it) quickly called on the Luftwaffe to participate in the great land campaigns, where its contribution was supremely effective. Economically, it is agreed that Germany's resources after it had taken over East-Central and Western Europe in 1938–1941 were greatly underutilized. Scholars now tend to explain this glaring inefficiency mainly by poor state coordination and rationalization of an economy managed by rival and divided party and bureaucratic 'estates'. Still, even if Germany decisively broadened its resource base, it practised a staged, 'stepping stone' scheme of expansion. Building on its relatively narrow, pre-1938 resource base to gain control over East-Central Europe, Germany then used its expanded power base to conquer Western Europe. Finally, the conquest of the Soviet Union's huge resources was intended to secure Germany economic autarky and lay the basis for its world power status.

Although Hitler was terrifyingly close to winning these successive series of gambles and achieving that goal, in the end he failed, as the Soviet Union survived and recovered.[17] So long as this goal was not achieved, and, in any case, certainly before the occupation of Western Europe in 1940, investment in heavy, 'strategic' bombers made no sense for Germany, either strategically or economically. Indeed such an investment – like the expansion of the navy – were only planned for c. 1944, after the Soviet Union would have been occupied and the struggle for world dominance would have begun. The reckless *va banque* gambling strategy of the 'stepping stone' model of expansion called for differently modulated armed forces and strict construction priorities in each successive stage until Germany broke away from its narrow resource base. It is true that in 1940–1942, the crucial stage of the war, Germany's utilization of the resources of occupied Europe was remarkably slack, a factor that effectively decided the war against it. Still, it should be kept in mind that even if Germany, as Overy claims, could have produced military hardware on a much larger scale if it had started to rationalize its war economy earlier than in 1942–1944, a limit would have ultimately been imposed on German production by the number of mechanized weapon systems that could be sustained on the Reich's meager oil resources. It was for this reason that Hitler was so desperate to get the Caucasian oil, whose incorporation was the only answer to the Reich's fundamental strategic dilemma.

All this goes to show that although official German air and armour doctrines in the mid-1930s were not very different from those prevalent in the other great powers, by the beginning of World War II they had each gravitated towards and found their place within a so-called 'Blitzkrieg' because of the particular aims and problems that shaped German strategy. These included the constraints imposed by Germany's war economy; Hitler's desire to break away from them once and for all; his 'stepping stone' expansion programme; and the widespread quest among the German military themselves for unconventional, revolutionary plans and techniques that would make possible swift decisions and avoid a protracted general war. That Blitzkrieg, for all its successes, proved brittle when encountering rivals with a wider resource base and more time to get their act together was not a failure of a military doctrine in the narrow sense; it was rooted in Germany's fundamental political-strategic impasse, which German strategy came frightfully close to overcoming but to which it ultimately succumbed.

The Soviet case offers instructive comparison and contrast to the German, because the Soviet Union's strategic conditions were in many ways similar to, but in many – very different from, those of Germany. The Soviet Union, too, was above all a continental power. It, too, was internationally isolated, because of her revisionist, revolutionary ideology, and it feared that a great power war coalition would form against it. It is this fear that dictated Soviet policy throughout the 1930s. However, unlike Germany, the Soviet Union was not constrained to act drastically. It is not only that Marxist ideology postulated that time was on the Soviet side in the long run, as the capitalist powers would collapse internally; more importantly, the Soviet Union also had practically all the resources needed to support a world power status and it was economically autarkic. Indeed, it was undergoing massive industrialization that was laying the ground for its future superpower status. In a technological-industrial form Soviet strategy continued Russia's traditional reliance on her immense space and resources. Although the Soviet economy and society – and consequently the armed forces as well – were still half-backward, these forces had at their disposal huge fleets of aircraft and land vehicles. At the same time, both strategically and ideologically, the Red Army was also committed to the mobilization of the masses in the nation in arms, only a small part of which could be mechanized.[18]

Hence the distinctive characteristics of the Soviet Deep Battle operational doctrine for the modern industrial-mechanized war. Ever since this doctrine has been discovered by Western observers from the 1980s on, it has been compared and contrasted with the German Blitzkrieg.[19] It has been rightly felt that for too long the former was overshadowed by the latter, for reasons that had little to do with their respective merits. By now, however, the pendulum seems to have swung too far in the other direction. The doctrine of Deep Battle, as it was crystallized in the early to mid-1930s, was outstanding in many ways. As mentioned earlier, it was the first comprehensive and highly advanced doctrine of the new mode of warfare officially formulated and instituted from above in any army. It was again unique in consciously integrating mechanized and non-mechanized forces within an overall operational concept. None of these were true of the would-be Blitzkrieg. Less favourably, the doctrine's insistence on the need for diverse tank types to match its elaborate mission definitions proved misguided.

All that aside, the claim that the Soviet Union's ultimate victory over Germany testifies to the superiority of Deep Battle over Blitzkrieg is either rhetorical or naïve. The war was not waged between two abstract doctrines but between two huge but very different industrial societies, and in the framework of a coalition war that pitted the resources of three of the world's four largest industrial economies against Germany. Indeed, it was precisely here that some of the crucial differences between Deep Battle and Blitzkrieg lay. Unlike the Germans, the Soviet strategists, drawing on the lessons of the First World War, envisaged and planned – could *afford* to plan – for a protracted total war that in view of the great industrial resources of the antagonists would not end in one stroke. Successive operations in time and space, leading from one campaign to another, would be necessary. Once Hitler's great gamble had failed – and it was not inevitable that it should have, for example, if the German war economy had been rationalized earlier, in 1940–1942, providing Germany with the extra equipment necessary for the destruction of the Soviet Union in 1941–1942 – it was indeed this sort of war that took over.

The creators of the Panzer arm took their inspiration from the pioneering British theory and practice in mechanized land warfare; this long-known fact was more recently called into question but is now firmly re-established. Less recognized is the fact that the Soviet authors of Deep Battle as well were crucially influenced by the British in adopting the ideas of mechanized warfare and deep operations at a critical stage of their development in 1928–1931. To be sure, they would adapt and develop these ideas in their own independent and highly innovative way.[20] But if, as was the case, all the armies of the industrial powers were keenly observing the British pioneering manoeuvres from 1927 on and reading the British theorists of mechanized warfare, why was it that from the mid-1930s Britain fell behind and never caught up in the development of mechanized forces and doctrine? This is a question that has intrigued students of war since the Second World War.[21] Unlike Germany, the Soviet Union, or, for that matter, France, Britain was not a continental power but mainly a maritime and colonial one. Regarding the First World War as a one-time episode, Britain demobilized its conscript armies after the war and recreated the small professional army, whose main mission was to police the empire. This mission was best carried out by foot or horse troops, or by lightly mechanized troops at most, rather than by the expensive 'robot army' proposed by the radicals. Thus, the British pioneering experimentation with mechanization barely translated into force formation and weapon acquisition.

From the mid-1930s, Britain was re-arming again to face the renewed threat of a great power war – but what sort of war? Unlike France (but like the United States), Britain placed its air and naval defence and deterrence forces first, at the expense of the army and a major overseas land involvement. Still, the strategic posture of the three Western liberal democratic great powers was in some fundamental ways similar, while again being radically different from that of Germany or the Soviet Union. Both Britain and France were status quo powers, and the United States shared with them a strong opposition to revision by war. This was not merely a function of a satisfaction with the status quo by the victors of World War I and possessors of large colonial empires, nor was it a function of capabilities alone, for the three liberal democratic powers differed widely among themselves in both respects. The United States, for instance, had no substantial imperial possessions, and its trade was hard hit by the creation of imperial economic blocks by Britain and France in the 1930s, during the Great Depression. The United States openly desired the abolition of protectionism and indeed the dissolution of the European colonial empires. It was also by far the world's mightiest industrial and, potentially, military power. Nonetheless, during the interwar period the United States never moved to realize that potential or attempted to redress its grievances by force, let alone establish a global military hegemony. In historical terms, the United States was a very strange 'hegemon' indeed. What truly distinguished the United States, Britain, and France, as opposed to the other great powers, was their identity as popular, liberal democratic, and individualistic societies. In the wake of the First World War these societies no longer regarded a serious great power war as an option that was compatible with their view of the world, well-being, and very low tolerance for the self-sacrifice of life in war. In this they anticipated the condition of the developed Western and Westernized countries ever since.

As the world was again becoming a dangerous place in the 1930s and the threat of a great power conflict was re-emerging, the Western liberal democracies were obliged to devise national policies and strategies to face the threats that would suit the kind of societies they had become. Indeed, though mostly unrecognized, it was in the field of strategic policy or grand strategy that the notions then developed in the West were of original and lasting significance. As we shall see in the following chapter in this volume, they would increasingly crystallize into a distinctive 'Liberal Way in Warfare' which is still with us today, whether nuclear or non-nuclear antagonists are confronted.

Notes

1 Azar Gat, *Fascist and Liberal Visions of War: Fuller, Liddell Hart, Douhet, and Other Modernists* (Oxford, 1998); incorporated with different pagination in my *A History of Military Thought: From the Enlightenment to the Cold War* (Oxford, 2001). (Pagination for both editions is provided here.)
2 Gat, *Fascist and Liberal Visions of War*, pp. 27–33; idem., *History of Military Thought*, pp. 545–51.
3 Heinz Guderian, *Panzer Leader* (London, 1952), p. 20.
4 Harold Winton, *To Change an Army: General Sir John Barnett-Stuart and British Armored Doctrine, 1927–1938* (Lawrence, KS, 1988).
5 See the first chapter in this volume.
6 Ibid.
7 Ibid.
8 William Mitchell, *Winged Defence* (New York, 1925), pp. 5–6, 99–138.
9 Winston Churchill, *The Second World War*, I (Boston, 1948), pp. 163–4.
10 E.g. Michael Geyer, *Aufrüstung oder Sicherheit. Die Reichswehr in der Krise der Machtpolitik* (Wiesbaden, 1980).
11 Gat, *Fascist and Liberal Visions of War*, pp. 80–103; idem., *History of Military Thought*, pp. 598–621.
12 For the debate see: A.S. Milward, *The German Economy in War* (London, 1965); Richard Overy, *War and Economy in the Third Reich* (Oxford, 1994); idem., 'Debate: Germany, "Domestic Crisis" and War in 1939,' *Past and Present*, 122 (1989), pp. 200–40. Adam Tooze's more recent *The Wages of Destruction: The Making and Breaking of the Nazi Economy* (New York, 2006), is to-date the best.
13 For this strangely belated but growing realization see Michael Geyer, 'German Strategy in the Age of Machine Warfare, 1914–1945,' in P. Paret (ed.), *Makers of Modern Strategy* (Princeton, 1986), pp. 585–6; Manfred Messerschmidt, 'The Political and Strategic Significance of Advances in Armament Technology: Developments in Germany and the "Strategy of Blitzkrieg",' in R. Ahmann, A. Birke, and M. Howard (eds.), *The Quest for Stability* (Oxford, 1993), pp. 249–61; J.P. Harris, 'The Myth of Blitzkrieg,' *War in History*, 2 (1995), pp. 335–52; and the first chapter in this volume.
14 Robert O'Neill, *The German Army and the Nazi Party* (London, 1966), p. 127.
15 Guderian, *Panzer Leader*, 29–30.

16 E. Homze, *Arming the Luftwaffe* (Lincoln, NE, 1976); Richard Overy, 'From Ural-bomber to Amerikabomber: The Luftwaffe and Strategic Bombing,' *Journal of Strategic Studies*, 1 (1978), pp. 154–78; Williamson Murray, *Luftwaffe* (Baltimore, 1985), pp. 4–15.
17 Andreas Hillgruber, *Germany and the Two World Wars* (Cambridge, MA, 1981); also Richard Overy, *Why the Allies Won* (London, 1995).
18 For the ideological aspect see my *Fascist and Liberal Visions of War*, pp. 114–21; *History of Military Thought*, pp. 632–9.
19 Richard Simpkin, *Deep Battle: The Brainchild of Marshal Tukhachevsky* (London, 1987); David Glantz, *Soviet Military Operational Art: In Pursuit of Deep Battle* (London, 1991).
20 See the first chapter in this volume; Gat, *Fascist and Liberal Visions of War*, pp. 116–9; idem., *History of Military Thought*, pp. 634–7.
21 See esp. Michael Howard, *The Continental Commitment* (London, 1972); Winton, *To Change an Army*.

3 Isolationism, appeasement, containment, limited war

The democracies' strategic policy from the modern to the 'postmodern' era

International relations theorists have become increasingly aware that modern democracies tend to avoid fighting one another, even though they continue to fight nondemocracies. Another noted aspect of liberal democracies' restraint is their tendency to eschew preventive war.[1] Historically, they have chosen not to initiate war even when they were threatened, held the military advantage, and were in danger of losing it. The British reluctance to seriously consider 'Copenhagening' the German navy during the massive German naval buildup prior to World War I is a case in point. The term alludes to Lord Nelson's pre-emptive destruction of the Danish navy in 1801 to prevent it from joining Napoleon. More strikingly, the liberal democracies did not intervene by force in the mid-1930s to prevent Hitler's Germany from re-arming, even though it meant they would lose the complete military superiority they held over Germany.

The United States demonstrated the same restraint vis-à-vis the Soviet challenge after World War II. It is generally forgotten that between 1945 and 1949, the United States possessed a monopoly over nuclear weapons. Theoretically, it had every reason to pre-empt without fear of retaliation rather than to adopt a strategy of containment and wait for Soviet nuclearization, which, although expected to come later than it actually did, was acknowledged as inevitable.[2] Theories of nuclear restraint often overlook this fact. But had it been the Soviet Union or Nazi Germany, rather than the United States, that possessed a nuclear monopoly, there can be little doubt that both would have pressed for the massive production of nuclear weapons and carried out a worldwide policy of conquest and coercion. Thus not only with other liberal democracies but also with its Soviet archrival, the United States refrained from pressing its overwhelming advantage to the point of war.

Less attention has been given to the question of *how* modern liberal democratic great powers have tended to respond to a threat and, if it has come to that, fight their wars. This chapter suggests that their strategic policy in facing this prospect has typically followed a pattern, progressing on an upward scale from isolationism to appeasement, containment and cold war, limited war, and only most reluctantly, to full-fledged war. As with other aspects of the liberal democracies' distinctive behaviour, this pattern began to manifest itself during the latter part of the nineteenth century, crystallized in the twentieth century, and is very much with us today.

What alternatives to war?

The notion that serious war is an unmitigated disaster and sheer madness increasingly took hold in the newly formed liberal democratic countries at the outset of the twentieth century, as the global industrial, trading, and financial system expanded (see the last chapter in this volume). It was against this background that World War I caused such a crisis in the liberal consciousness and traumatized liberal societies. Political scientist John Mueller has asserted that the trauma was a historical turning point that affected all powers alike, curing modern society of the predilection for war.[3] But this was not the case. The decline in belligerency did not start with World War I. On the contrary, the war came in the wake of the nineteenth century – from 1815–1914 – by far the most peaceful ever in the great power system. Nor was the deep trauma that developed in the aftermath of the war the result of the great losses of life and treasure in themselves. Again (as Mueller recognizes), these were not greater, relative to population and wealth, than the losses suffered in massive wars throughout history. The novelty was that liberal opinion now regarded such wars as wholly out of step with the modern world. The famous 'trauma' of the war was a liberal phenomenon, closely correlated with the strength of liberalism in each country rather than with that country's actual losses.

Britain, for example, was Europe's most liberal power, and the retrospective reaction against the war and the mourning for the 'lost generation' were the greatest there, even though Britain's losses were the smallest among the European powers. British casualties – three-fourths of a million dead – were terrible, but amounted to no more than 12 per cent of British troops enlisted during the war. They were smaller in absolute terms, and even more so relative to population, than France's loss of almost 1.5 million and Germany's 2 million dead. And yet the reaction against the war in Germany was far more limited than that in Britain.[4] Only liberal (and socialist) opinion, which was less dominant in Germany than in Britain, responded negatively. The most famous anti-war author was Erich Maria Remarque, a German liberal and pacifist. Certainly there was much war weariness and a widespread loss of enthusiasm for war in Germany, which, relative to its population, had suffered twice as many casualties as Britain. But Germany also had strong nationalist, anti-liberal, right-wing elements that vehemently opposed those anti-war sentiments. Ernst Juenger's books, glorifying his experience in the trenches and exalting the qualities of war, competed with Remarque's for popularity in Germany. Most importantly, powerful nostalgic evocation of soldierly trench camaraderie, embodied in voluntary paramilitary organizations such as the Blackshirts and Freikorps, played a major role in the postwar mood that helped bring to power, respectively, the fascists in Italy and the Nazis in Germany.

Perhaps the two cases that best illustrate the correlation between the post-World War I trauma and the level of liberalism are the United States and Serbia. The mightiest power in the world was not afflicted by heavy losses and crippling economic costs, as were the European belligerents. The United States suffered relatively light casualties in its brief involvement in the war and gained tremendously from the war materially, replacing Britain as the world's leading banker, creditor, and insurer. Nonetheless, it was in the United States that the onset of disgust with the war and regret over participating in it was the most rapid and sweeping. By comparison, the small and backward Serbia suffered, relative to population, the heaviest casualties of all the warring nations and was totally ravaged by the war and occupation. Nevertheless, Serbia hardly experienced the trauma of, and disillusionment with, the war. Nor, indeed, would other traditional societies that suffered hundreds of thousands and millions of casualties in the wars of the twentieth century (up to and including the Iran-Iraq War and Iraq's later gulf wars) react more traumatically than had been the norm among pre-industrial societies. By contrast, as the twentieth century ran its course, the smallest number of casualties has become sufficient to discredit a war in affluent liberal societies, particularly when the threat is not considered existential, imminent, or unsusceptible to alternatives to war, or when the prospects of achieving victory seem dim.[5]

Given the liberal democracies' fundamental attitude, the problem of how to deal with conflict has become a torment for them. Initially, liberals, while peacefully inclined, were not pacifist, because liberty had to be won and defended, even if by force. In time, some liberals (and socialists) came to espouse more or less unilateral pacifism. This, however, lacked a convincing explanation of what to do if the other side is not similarly pacifist, and so never became a dominant creed. More in tune with the liberal mainstream has been the effort to make the entire international system conform to the Painean-Kantian-Wilsonian model – that is, to have it embrace democratic self-determination, liberalism, and free trade, link into the modern spiral of mutual prosperity, and resolve disputes through international institutions. Where the conditions for that model materialized, as they did most notably in post-World War II Western Europe, the results were truly remarkable. But most of the world proved highly resistant to that model, and much of it still is.

Where a Painean-Kantian-Wilsonian peaceful accommodation fails to materialize, because not all states are liberal (and affluent), there remains the Abbé de Saint-Pierre's idea of collective security. In his *Projet pour rendre la paix perpetuelle en Europe* (1713), he suggested that all states combine against those that disturb the peace. This idea was central to the League of Nations and to the United Nations, but by and large it has failed for reasons long ago sensed by Rousseau:[6] powerful states and coalitions cannot easily be restrained by the threat of overwhelming collective action; the threat remains mostly theoretical, because states exhibit scant willingness to get involved in a conflict not their own; in the absence of a coercive authority that would prevent free-riding, states expect others that are more closely involved to do the job; states often have a greater interest in maintaining good relations with the aggressor; and determining who the aggressor is involves value judgements, about which no consensus can be reached. All this applies to democratic as well as to nondemocratic countries. The revived pleas in the post-Cold War era, and especially after 9/11, for collective security by consent through the United Nations greatly underestimate all these problems, as we have been witnessing in case after case.[7]

As long as the world has not become fully affluent, liberal, and democratic – and collective security remains largely ineffective – liberal democracies have been obliged to cope with the prospect of conflict and war. As already mentioned, their strategic policy in facing this prospect has typically followed a pattern, progressing on an upward scale from isolationism to appeasement, containment and cold war, limited war, and only most reluctantly, to full-fledged war. Thus the antagonist was to be disengaged if possible (isolationism); or, if its 'damage potential' was more considerable, its grievances were to be accommodated as far as possible and it was to be offered economic incentives to cooperate (appeasement); or, failing that, it would be subjected to deterrence, containment, economic sanctions, cold war, and limited military action at worst, mainly by naval, air, and elite land forces. All-out war was the very last resort.

Where isolationism/non-interference could be adopted, it has been a most tempting option for liberal democracies. However, in a shrinking world of growing interdependence, it has become increasingly untenable. Furthermore, even where no significant interests are involved, the liberal commitment to universal values and human rights often makes a foreign disturbance hard to ignore.

When faced with a significant threat that could not be shut out, the liberal democracies' option of second resort has been to compromise with a rival by accommodating some of its demands and offering it economic rewards. This option is cheaper than war, rests on the affluent liberal democracies' strongest asset – their abundant resources – and holds the prospect of integrating the rival into a mutually beneficial economic relationship that may eventually also lead to its liberalization. The success of such a policy of appeasement hinges on whether the other side chooses to accept the deal and become a partner, or views the offer as a sign of weakness that only whets its appetite. Thus states must appease from a position of strength and must dangle sticks in addition to carrots.

If appeasement fails, containment and cold war are the next steps in the sequence. These involve building a deterring coalition, applying economic pressure, and engaging in covert subversion and ideological warfare. Finally, if an armed conflict breaks out, liberal democracies attempt to limit its scope. They most often do this by providing money and hardware to cement coalitions and strengthen local forces against adversaries; employing blockades and naval and aerial actions, in which developed countries possess a clear superiority; and staging limited operations by technologically superior strike forces. Direct large-scale warfare, especially on land, where casualties might be high, has become the least desirable option. All these, of course, are 'ideal types' that often overlap. If they sound as if they come from today's headlines, their application in fact is of long standing.

The 'liberal way in conflict' – from 1900 to the present

During most of the nineteenth century, both Britain and the United States adopted isolationist policies. Britain was the first to be drawn out, when an external threat – most notably Germany's double challenge to Europe's continental balance of power and to Britain's naval supremacy – could no longer be contained without foreign commitments. Even then, however, Britain repeatedly sought in vain a combined naval, colonial, political, and economic deal that would bring about a *rapprochement* with Germany. And when war came, British policy was predicated on the assumption that most of the land war would be shouldered by France and Russia. Britain initially intended to confine its contribution primarily to the naval, amphibious, and economic spheres, with the blockade serving as its principal weapon. Only the imminent danger of its allies' collapse eventually forced Britain into full-scale military participation.[8]

The United States, for its part, was able to maintain its isolation for much longer. And even when he formally took his country to war in April 1917, President Woodrow Wilson did not plan full-scale involvement in the European carnage. The United States, too, was forced into full participation only by the near collapse of France and Italy and the spectre of British defeat in the U-boat campaign in the summer of 1917, the collapse and defection of Russia by the turn of the year, and the imminent disaster on the Western Front in spring 1918.

By the mid-1920s, the Western democracies' elites increasingly felt, in the spirit of John Maynard Keynes's *The Economic Consequences of the Peace* (1920), that the punitive Versailles treaty had been a mistake. During the Locarno Era they attempted to reach accommodation with Germany by helping it to revive its economy, by normalizing its international status, by integrating it into international institutions, and by holding before it the prospect of further peaceful settlement of its grievances. Unfortunately, this attempt collapsed with the post-1929 world economic crisis.

During the 1930s, the actions of Japan, Germany, and Italy against the international status quo posed acute threats to the liberal democracies. Nonetheless, in all the liberal great powers – the United States, Britain, and France – the public's mood and the consensus in the political parties and the government were unmistakably against war, even when the democracies still held the military advantage. Again, their policies evolved from isolationism to appeasement, to containment and cold war, to limited action. Total war was imposed on them by their enemies. All the liberal great powers trod that road.[9]

Isolationism was the preferred option of those who felt they could embrace it successfully: the British toyed with the idea and then adopted a policy of partial isolationism in the form of 'Limited Liability', which ruled out the commitment of substantial ground forces to continental Europe; the United States espoused isolationism more fully and for a longer period of time. However, given the magnitude of the threats, isolationism in itself was deemed to be insufficient. Both countries augmented the policy with efforts to ease the conflict and tame the Axis powers, especially Germany, by addressing some of their grievances and offering them economic rewards and beneficial trade deals. Most vigorously pursued by Neville Chamberlain, appeasement failed because ultimately Hitler's ambitions surpassed anything the liberal democracies were willing to accept. It should be noted, though, that those of Chamberlain's peers who opposed his policy did not object to appeasement as such, but believed it had to be more circumspect and buttressed by force.

During the Ethiopian and Spanish crises, the Western democracies still had little appetite for action in what they perceived as too small a threat. All the same, consider the type of strategies that were suggested (though mostly not implemented) to counter to the Axis moves. Rather than direct military intervention, they included economic sanctions, the isolation of both Ethiopia and Spain by the Allies' vastly superior naval power, and the supply of armaments to the Ethiopians and to the Spanish Republicans. In any event, it was only the Czechoslovak crisis in the spring of 1938 that greatly alarmed Western opinion. The strategic ideas pressed in opposition to appeasement again fit the pattern. In Britain, Anthony Eden, David Lloyd George, Winston Churchill, the British Labour and Liberal parties, and in the United States, President Franklin Roosevelt, all held that Germany had to be contained by a superior coalition (incorporating the Soviet Union), capable of deterring Germany or, failing that, of strangling it economically.

Roosevelt's line of thought with respect both to Europe and the Far East was typical. In late 1937, following Japan's invasion of China and the signing of the Anti-Comintern Pact among Germany, Italy, and Japan, the president increasingly aired the notion of a coordinated policy of sanctions and containment against the aggressors. The idea was embodied in his famous 'quarantine speech' of 5 December 1937. Later, during the Czechoslovak crisis, Roosevelt called for a 'siege' of Germany. He suggested that the European Allies close their borders with Germany, even without declaring war, and stand on the defence, relying on the economic blockade to do the job. The United States would back them economically.[10]

By the time war came in 1939, Germany had become less susceptible to economic pressure because of its domination of southeastern Europe and its pact with the Soviet Union. Under these circumstances, the 'Twilight' or 'Phony' War – 'Sitzkrieg' – being waged on the Western Front was not a curious aberration, as it is customarily regarded, but the most natural strategy for Britain and France. Having lost their ability to contain Germany within its old frontiers, choke it economically if it attempted to break out of them, or militarily defeat it and recover Eastern Europe from its grip, Britain and France in effect opted for more or less the same strategic policy that the West would adopt against the Soviet Union after World War II. They relied on armed coexistence, containment, economic pressure, and ideological and propaganda warfare. Militarily, Britain and France restricted themselves to peripheral and indirect action, trying to avoid escalation to full-fledged war. In all but name, this was a policy of containment and cold war. They hoped that over time, as the Western bloc formed its defences and deployed its resources, the Germans would be forced to seek an accommodation with the West. They also hoped that the Nazi regime might mellow or lose power. Unfortunately, the whole concept collapsed in May-June 1940, when Germany succeeded in decisively defeating the Allies and overran Western Europe.

The United States followed a similar path. In 1940–1941, American policy in both Europe and the Far East encompassed all means short of open war. A crucial element in Britain's decision to keep fighting in the summer of 1940 was Churchill's belief that the United States would enter the war before long, probably after the presidential election in November. This did not happen. Massive American economic aid in the form of Lend-Lease enabled Britain to continue the fight. But the prospect of an American declaration of war remained a dubious matter throughout 1941. During the summer of that year, the United States extended Lend-Lease to the Soviet Union, took over the battle against the German submarines in the western half of the Atlantic, and garrisoned Iceland. Nevertheless, it became clear to the British that American entry into the war was not to be expected in the near future. The majority of Americans and members of Congress objected to the war, and Roosevelt's own intentions were unclear. He was surely not going to allow Britain to fall, and probably would have used the United States' growing weight to steadily increase American influence on the course of the war. But was he waiting for more progress to be made on US rearmament, and using the time to prepare American public opinion for eventual participation in the war? Or was he quite satisfied with the existing situation, wherein Britain and the Soviet Union carried the burden of fighting, with massive American political and economic support but without full American participation? These questions remain in dispute and will probably never be answered conclusively. It is doubtful that Roosevelt himself knew. It was only Japan's surprise attack and the subsequent German declaration of war on the United States that decided the issue. Neither Britain nor the United States embarked on all-out war until forced into doing so by the surprising collapse of their defences, in May–June 1940 in Western Europe, and in December 1941 in the Pacific, respectively.

Although far more powerful than Japan in all respects, the United States deployed nonmilitary means to contain Japan in 1940–1941. Washington tightened economic sanctions so strongly that the imposition of an oil embargo threatened to bring Japan to its knees. Unfortunately, defensive precautions to back up this policy proved insufficient. As it had with Germany the year before, the policy of containment, economic coercion, and cold war floundered when the enemy did the unthinkable, and in a highly successful lightning campaign, broke down the walls that had been built up against it.

By the end of World War II, the Soviet Union was taking the place of the Axis powers as the liberal democracies' major potential rival. And, yet again, the democracies' response followed a path leading from appeasement to containment and cold war. As revisionist historians of the 1930s have reminded us, towards the end of the war, Roosevelt and Churchill recognized Soviet control over Eastern Europe for much the same reasons that Chamberlain had been prepared to see German hegemony extended over that region. Roosevelt in particular hoped to come to terms with the Soviet Union and incorporate it within a new global four-power collective security system. By 1946–1947, however, American hopes were dashed, and the policy of containment and the Cold War came into being.

As noted previously, this policy was adopted when the United States still held the monopoly over nuclear power. All the same, according to George Kennan, the intellectual architect of containment, the idea was formed in a fundamentally nonnuclear frame of mind and derived from pre-1945 experiences.[11] The atomic bomb is not even mentioned in either his 'Long Telegram' from Moscow in February 1946 or his famous 'Mr. X' article in *Foreign Affairs* in 1947, containment's formative documents. Throughout the second half of the 1940s, Kennan insisted that the United States must refrain from using nuclear weapons as an active instrument of diplomacy and war. Periods of heightened tensions and greater militarization alternated with periods of *rapprochement* and *détente* until the end of the Cold War.

In the nuclear age, the prospect of a major great power war diminished, and it appears to have become even more remote since the collapse of the Soviet Union and the communist challenge. More tellingly, with the exception of the preventive war in Iraq, carried out in the special circumstances of the 'War on Terror', the same pattern has manifested itself even in liberal democracies' conflicts with far smaller and weaker, economically backward, rivals. Most typically, the pattern has been unfolding in the effort to stop and reverse the nuclearization of North Korea and Iran. The full range of options – from appeasement, to containment, cold war, and limited war – has been practised or entertained by the United States and its democratic allies. In both cases it remains to be seen how things will unfold.

This pattern of conflict behaviour has had a mixed and often disappointing record since it began to crystallize in the early twentieth century. The methods involved are politically and strategically difficult to apply, often ineffective, and carry their own psychological strains. Furthermore, there is a catch that lies at the root of affluent liberal democracies' torment in conflict situations. Since wars are abhorred in liberal societies as antithetical to both their interests and values, to their entire way of life and worldview, they are sanctioned only as a last resort – after all other options have failed. Yet in practically no situation does it ever become clear that all alternative policies have indeed been exhausted, that war has really become unavoidable. A feeling that there may be another way, that there *must* be another way, always lingers on. Errors of omission or commission are ever suspected as being the cause of undesired belligerency. Moreover, it never becomes clear that the democracies come to a conflict with entirely clean hands, morally, because of past or more immediate alleged wrongs; nor indeed can they, given the inevitable gap that always separates ideals from reality. All in all, given the nature of modern liberal societies, the previously described way of coping with a threat and making war appears to be their norm.

Notes

1 Randall Schweller, 'Domestic Structure and Preventive War: Are Democracies More Pacific?' *World Politics*, 44 (1992), pp. 235–69.
2 John Gaddis, 'The Origins of Self-Deterrence: The United States and the Non-Use of Nuclear Weapons, 1945–1958,' in John Gaddis (ed.), *The Long Peace: Inquiries into the History of the Cold War* (New York: Oxford UP, 1987), pp. 104–46.
3 John Mueller, *Retreat from Doomsday: The Obsolescence of Major War* (New York: Basic Books, 1989).
4 *Contra* Ibid., pp. 53–68.
5 Eric Larson, *Casualties and Consensus: The Historical Role of Casualties in Domestic Support of US Military Operations* (Santa Monica, CA: Rand, 1996); C. Gelpi, P. Feaver, and J. Reifler, 'Success Matters: Casualty Sensitivity and the War in Iraq,' *International Security*, 30:3 (2005–2006), pp. 7–46; idem., 'How Many Casualties Will Americans Tolerate?' *Foreign Affairs*, 85:1 (2006), have claimed that American casualty sensitivity is overrated. They demonstrate that it was actually a function of the public's assessment of the prospects of achieving military victory. However, nondemocratic and

less developed countries are less susceptible to this constraint, and their casualty tolerance in general is much higher. Moreover, liberal democracies' very ability to achieve victory in some types of war has been severely constrained by their self-imposed norms of restraint (see chapter 7 in this volume), which takes us back to square one: liberal democracies are more casualty-sensitive because they are liberal and democratic.

6 Jean-Jacques Rousseau, 'Abstract and Judgement of Saint-Pierre's Project for Perpetual Peace' (1756), in S. Hoffmann and D. Fidler (eds.), *Rousseau on International Relations* (Oxford: Oxford UP, 1991), pp. 53–100.

7 All this is insufficiently addressed in Richard Rosecrance (ed.), *The New Great Power Coalition: Toward a World Concert of Nations* (Lanham, MD: Rowman, 2001).

8 David French, *British Strategy and War Aims, 1914–1916* (London: Allen and Unwin, 1986).

9 This is the subject of part II of Azar Gat, *Fascist and Liberal Visions of War* (Oxford: Oxford UP, 1998); incorporated in Azar Gat, *A History of Military Thought* (Oxford: Oxford UP, 2001).

10 David Reynolds, *The Creation of the Anglo-American Alliance, 1937–1941* (London: Europa, 1981), esp. pp. 17, 30–1, 35; Robert Dallek, *Franklin D. Roosevelt and American Foreign Policy, 1932–1945* (New York: Oxford UP, 1979), especially pp. 163–4; Callum Macdonald, 'Deterrence Diplomacy: Roosevelt and the Containment of Germany, 1938–1940,' in R. Boyce and E. Robertson (eds.), *Paths to War: New Essays on the Origins of the Second World War* (London: Macmillan, 1989), pp. 297–329; D.C. Watt, *Succeeding John Bull: America in Britain's Place, 1900–1975* (Cambridge, England: Cambridge UP, 1984), pp. 82–3.

11 George Kennan, *American Diplomacy* (Chicago: U. of Chicago P., 1985 [1951]), pp. vi–vii.

4 The 'Revolution in Military Affairs' (RMA) compared with earlier military-technological revolutions of the nineteenth and twentieth centuries

The conspicuous changes that have taken place in the face of warfare over the past decades have been titled the 'Revolution in Military Affairs' (RMA). The problem with this label, however, is that it tells us nothing about the nature of the revolution and its place in the broader context of technology-driven revolutions of the industrial-technological age. These have been nothing short of the defining developments of modernity. Over the past two centuries, innovation in technology accelerated dramatically in comparison to pre-industrial times, with military technology constituting merely one aspect of a general trend.

In premodern times, too, technology mattered, and some innovations in military technology profoundly affected warfare, and history in general. Metal weapons, equestrian technology, the longbow, rowing and sailing ships, and firearms are often-cited examples, and there are many more. And yet, technology improved slowly in the premodern era, so that something close to equilibrium often prevailed for centuries and millennia between each significant 'punctuation' in the evolution of military technology. The main infantry weapon, the musket, changed little from 1690–1820. However, from the beginning of the industrial-technological era, as military theorist J.F.C. Fuller saw, the pace of technological innovation became such that the best armed force of one generation would have been totally unable to confront in the open a well-equipped opponent of the following generation.

As Fuller equally saw, the advances in military technology were closely related to civilian developments; and rather than taking place more or less evenly in time and across the technological front, both mainly clustered around consecutive breakthroughs in a number of sectors each time.[1] Taking decades, indeed several generations, to run their course, these technological breakthroughs then gave way to other breakthroughs in different sectors. Although some oversimplification is necessarily involved, Fuller rightly identified three such major revolutionary waves of civil-military technological change during the nineteenth and twentieth centuries.

Three revolutionary technological waves

The so-called First Industrial Revolution, taking about a century to unravel, centred on the steam engine and on major advances in metallurgy and machine tools. The steam engine, practically the only engine in existence until the late nineteenth century, was applied to propel all sorts of different machines, revolutionizing one field of human activity after another. Originally developed to pump out water from mines, it was then harnessed to the newly-developed spinning and weaving machines of the cotton industry, revolutionizing textile production. Applied to pull trains of wheeled carriages that ran on railroads, it revolutionized land transportation from the 1820s on, placing it for the first time on equal footing with water transportation and opening up the interior of the world's great continental land-masses. And harnessed to a paddlewheel and then a propeller, it finally displaced the great sailing ships, one of the pinnacles of pre-mechanized technology. All these were various applications of the same basic technology.

Those changes affected the military field as deeply as they did civilian life. The railway increased armies' strategic mobility and logistical capability by a factor of hundreds. While naval mobility only doubled or tripled as steam replaced sail, naval tonnage grew fourfold to fivefold and (iron and steel) battleship's size – and might – tenfold and more.[2] To these was added the revolution in information communications, as electric telegraph lines connected not only armies across countries but naval bases across oceans and continents on real time, where weeks, months, and years had been necessary.

Simultaneously during the nineteenth century, the revolution in metallurgy (iron followed by steel) and machine tools generated a revolution in firearms and tactics. Rifling and breech-loading were pioneered in infantry firearms during the 1840s, and in artillery during the 1850s and 1860s. Magazine-fed rifles, 'repeaters', were developed in the 1860s and 1870s, and quick-firing artillery, using a hydraulic mechanism to absorb the gun's recoil, in the 1880s and 1890s. In consequence, range, accuracy, and rapidity of fire each increased some tenfold within 60 years, not counting the development of the automatic machine gun from the 1880s, which multiplied firepower yet more.[3] Naval gunnery underwent similar developments, to which the torpedo was added from the 1870s.

All these, however, were lopsided revolutions, especially on land. As in the economy so in the military: spheres of activity to which the steam engine could not be applied remained manual and unaffected by the Revolution. Thus, while armies rode trains on their way to the battlefield and were easily controlled by telegraph, they fell from the pinnacle of high-tech communications back to Napoleonic if not Alexandrian times once *on* the battlefield. Their campaign and tactical mobility remained confined to human muscles, with their artillery and supply drawn by horses. Hundreds of thousands and millions of horses remained in each of the great powers' armies during World War I, and in some, including the mythically mechanized German army, also throughout World War II. Field command and control, where telegraph lines could not be laid in advance, was similarly downgraded to messengers on foot or horseback. Furthermore, whereas firepower increased tenfold and more, troops, while dispersing and taking cover, still had nothing better than their skin to protect them from the storm of steel on the open field. Hence the murderous stalemate on the Western Front during World War I, both tactical and operational. Even those puny gains made by attacking infantry at terrific cost were reversed as decimated foot-walking troops, struggling to extend their tactical gains deeper, were pushed back by enemy reinforcements rushed up by rail.

However, from the 1880s on a new revolutionary wave of industrial technology, the so-called Second Industrial Revolution, was beginning to unfold in civilian life, affecting the military field as profoundly as the First Industrial Revolution had. Chemicals, electric power, and the internal combustion engine dominated that second revolutionary wave. While the chemical industry contributed high explosives – remember Alfred Nobel – and was soon to produce chemical warfare, and while developments in electricity also had various military applications, including radio communication, it was the internal combustion engine that affected war the most decisively. Lighter and more flexible than the steam engine, it made possible mobility in the open country, away from railways. Passenger and transport automobiles (as well as the tractor) rapidly evolved between 1895 and 1905, increasing cross-country mobility by a factor of tens. World War I inaugurated the tank – an armoured and armed tractor – which introduced mechanized mobility and mechanized armoured protection into the battlefield, thereby redressing the huge imbalance created by steam. Controlled by radio, which similarly extended real-time information communication into the open field, away from stationary telegraph lines, mechanized armies on tracks and wheels matured by World War II, some half a century after the pioneering of the technologies that had made them possible.

Simultaneously, the internal combustion engine also made possible mechanized air flight. A remarkably similar trajectory followed, with the first such flight taking place in 1903, and massive air forces quickly coming into being during World War I and further developing by World War II. Ships, already steam-powered and armoured, were less dramatically affected by the internal combustion engine. Nonetheless, naval warfare in general was revolutionized. Dual propulsion by the internal combustion and electric engines made possible the first workable submarine in 1900, while the aircraft was to bring about the demise of the gunned battleship. Thus, the automobile, the submarine, and the aircraft all made their appearance in close proximity roughly between 1895 and 1905, made possible by the same new technology. They all developed by leaps and bounds during World War I, and together they completely dominated both land and naval warfare in World War II.

By then new technological breakthroughs were beginning to make their mark in other sectors, most notably electronics, again revolutionizing both civilian life and war in the so-called Third Industrial or Information Revolution. Radar, developed in the late 1930s, deeply affected air, air-land, and sea warfare in the following decades. From around 1970, electro-optic, television, and laser guidance for missile weapon systems began to revolutionize air-land and land battle. Since then, sensors of all sorts have been rapidly improving, in connection with the fast miniaturizing microchip that has doubled electronic computation capacity every 18 months for nearly half a century. The chip – the steam or internal combustion engine of our times – has been applied to a dazzling array of new technologies, revolutionizing each. As a result of all this, the identification, acquisition, and destruction of most hardware targets have become almost a foregone conclusion, nearly irrespective of range. Showing little signs of levelling off, the electronic revolution is bringing about increasing automation. This is the electric-robotic warfare that the pioneering Fuller predicted as early as 1928 as the third great wave after mechanization.[4]

The far-reaching effects of the ongoing electronic-information revolution on warfare have been endlessly discussed. The old mechanized armoured armies of the previous era may not be disappearing, but they have been shrinking in size and transformed to embrace electronic warfare themselves, defensively as well as offensively. The two Gulf wars demonstrated this most strikingly, for the Iraqi side that lacked the new technologies found out to its cost that its numerous old-style formations were as vulnerable as herds of prehistoric mammoths. The gap between developed and less developed protagonists seems to have widened considerably. And yet the latter have been adjusting quicker and in ways different than expected.

In the first place, less developed players have been moving to get rid of their heavy formations, adopting instead low-signature troops, weapons, and tactics. They aim to slip under the threshold of the electronic weapon systems, which are much better at identifying hardware than people. (Notably, though, the weaker side cannot dispense of its heavy dual-purpose civilian-military infrastructure that remains highly vulnerable, as Serbia's experience in the Kosovo War demonstrated.) Second, the massive market penetration of new technologies into every aspect of daily life makes them available to less developed players as well, if not in the form of the most expensive cutting-edge military systems then as widely available and cheap gadgets. Satellite navigation systems (GPS) that offer precision guidance, computer networks that can be exploited and disrupted, and cellular phones that can be activated from afar, are some examples. Indeed, high-tech technologies have both polarized *and* democratized the balance between the more and less advanced sides in war, for the means to generate massive damage with pinpoint accuracy have been trickling down to below the state level, becoming available to non-state actors as well.

Common traits and fundamental problems

We now progress from the survey of history to discuss some general, recurring traits of military-technological revolutions.

Force multipliers, one-sided battlefield outcomes, and cancelling-out effects

With arms races gaining a wholly new significance in the technological age, swift changes in, and redresses of, the balance of power due to innovation have become very much the rule. Particularly when one side succeeds in securing a decisive lead in the acquisition and assimilation of breakthrough weapon systems, it might gain a 'force multiplier' that can produce one-sided battlefield results. Examples include the Prussian breech-loading rifle in the 1866 war against Austria, the German mechanized forces at the beginning of World War II, and the American electronic weapon systems in the campaigns of the past two decades. Over time, however, there would be a canceling-out effect, as rivals catch up in the development and assimilation of new weapon systems and operational doctrines, and adopt countermeasures. This, too, was early pointed out by Fuller. Breech-loaders responded to breech-loaders; mechanized forces and anti-tank weapons countered mechanized forces; electronic detection, disruption, and suppression devices target electronic weapon systems.

Technology could also be transferred – sold or given – which occasionally narrowed the gap in power between more and less advanced rivals. Finally, extreme technological asymmetry has been countered by asymmetrical strategy. The Battle of Omdurman (1898) in the Sudan, where Kitchener's British army, employing machine guns and magazine-fed rifles, annihilated the Dervish army, killing 11,000 at the price of 140 British casualties, demonstrated the huge gap that had been opening between the armies of technologically advanced powers and their rivals from less developed societies. Thereafter, the latter tended to avoid direct fighting, opting for guerrilla and other methods of irregular warfare. At sea and in the air, where fighting has been the most hardware-intensive and has been carried out entirely by weapon systems (rather than by armed men), the gap between technologically more and less advanced rivals has been the widest.

Growing lethality?

Intertwined as it has been with the general development of technology, weapon systems' effectiveness, while difficult to measure exactly, probably increased at the same rate as productivity in general during the industrial-technological age: that is, by a factor of hundreds. But what exactly does this increase signify? Destructiveness or lethality may appear to be the issue, for that is what war is all about. On the other hand, as the prior overview has already demonstrated, developments in military technology also exponentially increased protective power – for example, through mechanized defensive armour at sea and on land, through growing, indeed, sometimes literally rocketing, swiftness and agility, and through electronic counter-warfare. Contrary to widespread assumptions, studies of war lethality, measured by military and civilian casualties, show no significant increase during the nineteenth and twentieth centuries in comparison to earlier historical periods, relative to population.[5] Mortality as percentage of the armed forces did not grow. Furthermore, the vast majority of the non-combatants killed by Germany during World War II, the most severe modern war – Jews, Soviet prisoners of war, Polish and Soviet civilians (millions in each category) – fell victim to intentional starvation, disease, and mass executions, rather than to any sophisticated military technology. Indeed, similar and greater rates of combat mortality than that experienced in World War II had been commonplace among pre-state, small-scale societies that used the most primitive military technology. Instances of genocide in general during the twentieth century, as earlier in history, were carried out with the simplest of technologies, with Rwanda being a chilling recent example.

Spiraling costs?

Another general characteristic of the industrial-technological age with its consecutive military revolutions has been the steadily growing weight of hardware in military spending. However, again contrary to a widely held view, weapon systems' cost, like military spending in general, did not spiral out of control with every new generation, nor did they become prohibitively expensive. Military spending does not simply rise with a 'free-floating' cost of weapons, but is always bounded by economic capabilities, by priorities, and by enemy's spending bounded by similar constraints. In fact, the peacetime military spending of the powers during the industrial age has remained remarkably stable, at less than 5 per cent of GDP. That was the level of military spending in Britain, the leading industrial power of the nineteenth century, as well as in the other European great powers. Although registering some increase, military spending retained these modest levels even during the arms race between the European powers in the decade before World War I.[6] During the twentieth century, peacetime military spending generally retained the same, surprisingly resilient, levels as in earlier centuries: over 5 per cent of GDP for the United States during the Cold War; around 3 per cent for the European countries. Thus, military spending did not 'spiral', but generally kept in line with the overall growth in GDP. Only in wartime, most notably during the two world wars, did the military spending of the warring great powers leap to around half of total GDP.

What is true, however, is that hardware's *share* in military spending has grown during the industrial-technological era, with the rising cost of some weapon systems compensated for by smaller numbers or by a changing balance within military procurement, both matters of cost-effectiveness.[7] The problems of priorities that the American defence authorities are facing today with such expensive weapon systems as the air-superiority aircraft F-22 Raptor and F-35 Lightning II are not new.

During the early modern period pay and provisions for the men far outweighed the cost of military hardware, not only in the armies but also in the much more capital-intensive navies. However, as the armed forces became increasingly mechanized and hardware-dependent, this ratio tilted in the other direction. In the navies, the first intensively mechanized service, manpower costs (including pay, provisions, and clothing) declined to about 40 per cent of the naval budgets by the last decade of the nineteenth century, and to around 30 per cent by the first decade of the twentieth century. At the same time, warship construction costs grew to over one-third of the naval budgets, with nearly as much additional spending dedicated to repairs, maintenance, and ammunition. Thus materiel grew to outweigh manpower costs by 2:1.[8] The US Navy's budgetary figures a century later, in the 2000s, remain in the same range for manpower, procurement plus R&D, and operations and maintenance, respectively.[9]

Armies were far less mechanized than navies around 1900, so manpower costs (direct and indirect, where conscription was involved) undoubtedly still constituted the majority of army expenditure in all the great powers. Yet this picture is misleading, because one of the major elements of military power was the railroad system – for mobilization, strategic mobility, and logistics. And the overwhelmingly civilian infrastructural costs of that system (and others) did not register in military budgets. From World War I on, however, armies increasingly mechanized. Consequently, manpower costs in the US Army, for example, fell to around 40 per cent of the budget in the 2000s, even though this is an all-professional rather than a conscript force. Materiel costs (excluding construction) amount to about half of the budget, with operations and maintenance costing double as much as new procurement and R&D.[10]

Finally, while armies remain the most labor-intensive armed service, air forces are the most capital-intensive of all armed services. Manpower costs amount to only 20 per cent of the US Air Force's budget in the 2000s, with operations and maintenance, procurement, and R&D together taking up about three-quarters of the budget.[11]

Prediction under conditions of discontinuous experience

The fundamental challenge posed by military-technological revolutions is that they involve a sweeping change that goes far beyond existing experience. The contours of the change are shrouded in the mists of the future, and can only be more or less successfully predicted. Nonetheless, huge investments in hardware and far-reaching adjustments in military structure and operational doctrine, which can only be tested in battle, are called for in the present. To be sure, the uncertainty of the future is not confined to the military sphere. Firms, states, and individuals are required to make decisions on future actions and investments based on predictions. Still, the military sphere is special in that the practical experience of war is not continuous but is usually limited to relatively short spates of active belligerency, separated by long periods of peace. Although armed forces attempt to overcome these gaps by means of operational analyses, simulations, and exercises of all sorts, a continuous reality-check for the military does not exist in the same way that it does, say, in economic life. Granted, the limitations of reality-check in the economy, too, have recently been most poignantly demonstrated by the buildup, and crash in 2007–2008, of the credit bubble economy, with the mountain of economic assumptions and expectations that went with it. But the limitations are even more severe where experience is not continuous.

In Europe, for example, peace prevailed among the great powers for 43 years, from 1871–1914. Wars took place only on the peripheries of the system. As a result, even though the exponential rise in firepower during that period was well recognized by all armies, its significance was not fully assimilated until the actual test of fire in World War I. The long interruptions in the experience of war is the reason why armed forces are proverbially inclined to prepare for the last, rather than the next, war, a problem that, indeed, became most acute only with the revolutionary technological changes of the industrial age.

I assume most of us would have gladly traded combat experience for peace. Unfortunately, though, during the early twenty-first century the American armed forces, and the Israeli, have gained continuous combat experience, with the result that gaps in the assessment of the use of modern weapon systems on the electronic battlefield narrow, and continuous adjustments are carried out. Much of this experience, however, is limited to counterinsurgency warfare and to regular combat against far weaker rivals. And limited experience in specialized circumstances can be more misleading than no experience at all. The lessons of the Spanish Civil War, for example, retracted the development of independent mechanized forces, most notably in the Soviet Union.

Grand visions and practical designs

It may not be redundant to point out that the problem in dealing with technological revolutions is not so much to realize generally that such a revolution is in fact occurring, but to work out its exact form and practical implications. While inspiration is crucial, God has long been proclaimed to be in the details. Insufficient awareness of a brewing revolution may exist in its formative stage; but awareness grows quickly as the revolution is gathering momentum, and all the more so in our technological society, which has grown accustomed to such changes and experiences them in all spheres of life. The real problem is to identify the precise nature of the expected changes in war, devise concrete programmes of transformation in the organization and doctrine of the armed forces, and determine the optimal mix of hardware to be developed and procured. All these complex and critical questions are interconnected, and they are saturated with technological and other unknowns, not to mention budgetary and operational constrains. There were many early car manufacturers, but only Henry Ford got it exactly right, at least for a while. There have been thousands upon thousands of high-tech companies, but only the Microsofts, Apples, Googles and Facebooks developed winning combinations.

During the 1930s, for example, contrary to mythology, the debate within the armed forces did not take place between far-sighted exponents of mechanization and their reactionary rivals, but revolved around far more practical questions: should all tanks be concentrated in armoured divisions, or should part of them be allocated to infantry support? And in connection with this, what types of tanks should be manufactured, and what would be the role of infantry and other traditional arms on the modern battlefield?[12] Conclusive answers to these questions were only given in the practical test of World War II, and not always along the lines suggested by the radicals.

This applied even more to air forces. Their major role in any future war was recognized by all, and it was in fact the radicals who proved mistaken in their most central assumptions. They believed that air forces were capable of easily and swiftly destroying the industrial infrastructure of great powers and force them to capitulate. They thus contended that armies and navies had only a secondary role to play, if they had not become wholly obsolete. At sea, it was not clear, for example, if the invention of ASDIC/sonar did not spell the end of the submarine, as Winston Churchill, for one, hoped. Nor was it known if the aircraft had rendered the battleship obsolete or would coexist with it. These critical questions were tested, and settled, only in World War II.

The nuclear and biotechnological revolution

Our discussion so far has been confined to conventional warfare. Obviously, though, by far the most revolutionary technological breakthrough of the modern era has been the advent of nuclear power and nuclear weapons. The overwhelming destructive power of a nuclear explosion is offset by no parallel rise in defensive power, as has been the case with other military technologies. This is finally the ultimate weapon, a doomsday machine that gives anybody who possesses the necessary stockpile an assured ability to completely destroy his enemy – indeed, the world. At the same time, the fact that no effective defence against nuclear weapons exists means that when both sides possess them their actual use in war has been all but prevented by deterrence of supreme potency: the promise of mutually assured destruction (MAD). This was the logic that dominated the Cold War and served as the ultimate break against it becoming hot. And yet this logic may no longer apply as comprehensively as it did.

The change is driven in part by technological innovation and in part by technological proliferation. It concerns not only nuclear but also biological weapons. Nuclear technology has been around since 1945. But as both its civilian and military uses are becoming more widely available, the Non-Proliferation Treaty regime built to contain it is in a real danger of collapse. Biological weapons have also been around since before World War II. Yet with the breakthroughs achieved in decoding the genome and genetic engineering, the biotechnological sector has become one of the spearheads of today's scientific-commercial revolutionary wave, together with electronics. As a result, biological weapons have become both much more lethal and accessible. A virulent laboratory-cultivated strain of bacteria or virus, let alone a specially-engineered 'superbug' with no immunological or medical antidote,[13] could make biological weapons as lethal as nuclear attacks. At the same time, they are far more easily available than nuclear weapons.

By the late 1990s, there were already an estimated 1,300 biotechnology companies in the United States and 580 in Europe.[14] Since then, the number of scientists authorized to work with biological agents has risen to 15,000 in the United States alone, rendering monitoring by government agencies all the more difficult.[15] One estimate suggests there are some 20,000 labs in the world, where, within the next decade, a single person will be able to synthesize any existing virus. In the same labs, five people with 2 million USD will be able to create an enhanced pathogen – a virus that could infect and kill people who have been immunized with conventional vaccines. With 5 million USD, the same five people could build a lab from scratch, using equipment purchased online.[16]

The root problem in this process is the trickling down to below the state level of the technologies and materials of WMD. The ability of non-state actors to buy, steal, rob, and/or manufacture WMD has increased dramatically. Here lies the bewildering nature of the new mix which is unconventional terror. Not only are terror groups more likely to consist of zealots who are willing to sacrifice their lives and may positively desire a general apocalypse; they are also too elusive to offer a clear enough target for retaliation, on which the whole concept of deterrence is based. For this reason the use of ultimate weapons is *more* likely to come from them than it is from states, even though the latter possess far greater capability. In contrast to the mindset that dominated strategic thinking since the onset of the nuclear age, unconventional capability acquired by terrorists is *useable*.

The 'encapsulation' of destructive power in WMD, particularly the nuclear and biological, creates a situation whereby one no longer has to be big in order to deliver a devastating punch. Scenarios of world-threatening individuals and organizations, previously reserved to fiction of the James Bond genre, suddenly become real. The greatest threat of nuclear and biological proliferation into countries with low security standards and high levels of corruption is not so much that of actual military use by the states involved but the much increased danger of leakage to non-states. Not only might people and organizations with access to nuclear facilities sell or otherwise transfer nuclear materials, expertise, and even weapons to others; states in the less developed and unstable parts of the world are also ever in danger of disintegration and anarchy. Abdul Qadeer Khan, the Pakistani nuclear scientist who headed his country's programme to manufacture an atomic bomb, sold the nuclear secrets to perhaps a dozen countries, reportedly for as little as millions or tens of millions of dollars. And the prospect of a Taliban takeover, or general anarchy, in Pakistan is looming as these lines are written, to the concern of the world community. The disintegration of the Soviet Union left in the debris of its advanced military infrastructure unemployed scientists, poorly guarded production facilities, unaccounted for materials, and, most troubling of all, the weapons themselves. Indeed, the collapsed Soviet Union rather than the former nuclear superpower may be the model for future threats. In vast and barely approachable tracts of the globe inhabited by fragmented and unruly societies, the ability to monitor unconventional weapon production or acquisition is inherently limited. The difficulties of finding a needle in a haystack pale in comparison.

According to Rolf Mowatt-Larssen, the US Energy Department's former director of intelligence, al-Qaeda was seeking to acquire weapons of mass destruction for nearly a decade before September 11. Osama bin Laden offered 1.5 million USD to buy uranium for a nuclear device. In August 2001, just before the attack on the United States, bin Laden and his deputy, Ayman al-Zawahiri, met with Pakistani scientists to discuss how al-Qaeda could build such a device. Al-Qaeda also had an aggressive anthrax programme that was discovered in December 2001, after the organization was driven from Afghanistan. In 2003, Saudi operatives of al-Qaeda tried to buy three Russian nuclear bombs. Around the same time, Zawahiri decided to cancel a cyanide attack in the New York subway system, telling the plotters to stand down because 'we have something better in mind'. After 2004, al-Qaeda's WMD trail went cold.[17] The price of al-Qaeda's successful terror attack on 9/11 may have been to nip its far more dangerous plans in the bud. It seems clear that the loss of the Afghani safe haven was a serious blow to al-Qaeda's WMD programme, though, of course, the price of success is that we shall probably never know for certain. The failure of al-Qaeda to initiate further terrorist acts on US soil after 9/11 (pointed out by those who believe that the danger has been exaggerated) is largely attributable to the loss of that safe haven. As with other effective countermeasures, the paradoxical conspicuousness of the 'dog that didn't bark', in Sherlock Holmes's words, should not become a cause for complacency.

Notes

1 These ideas repeatedly occur in J.F.C. Fuller's voluminous writings; but see esp. his *On Future Warfare* (London: Praed, 1928); idem., *Armament and History* (London: Eyre, 1946).

2 Data for Britain. See Quincy Wright, *A Study of War* (Chicago: U. of Chicago P., 1965), pp. 670–1 (military); B.R. Mitchell, *International Historical Statistics, Europe, 1750– 1988* (New York: Stockton, 1992), tables F4 (merchant).

3 Of the many references to these developments, Dennis Showalter, *Rifles and Railroads: Soldiers, Technology and the Unification of Germany* (Hamden, CT: Archon, 1975); Daniel Headrick, *The Tools of Empire: Technology and European Imperialism in the Nineteenth Century* (New York: Oxford UP, 1981), are the most expert.

4 J.F.C. Fuller, *Towards Armageddon* (London: Dickson, 1937), pp. 92, 132.

5 See the last chapter in this volume. [The general trend is more or less agreed upon by statistical studies such as: Melvin Small and David Singer, *Resort to Arms: International and Civil Wars, 1816–1980* (Beverly Hills: Sage, 1982); Jack Levy, *War in the Modern Great Power System, 1495–1975* (Lexington: UP of Kentucky, 1983); Evan Luard, *War in International Society* (London: Tauris, 1986); Niall Ferguson, *The Cash Nexus: Money and Power in the Modern World, 1700–2000* (New York: Basic Books, 2001), pp. 33–6, makes the point for spiraling lethality, but he scarcely controls for population (also as a function of the geographical scope of wars) and mobilization levels, and appears to be unaware of the heightened protective aspect of improved military technology.]

6 David Stevenson, *Armament and the Coming of War: Europe, 1904–1914* (Oxford: Oxford UP, 1996).

7 Cf. Ferguson, *The Cash Nexus*, pp. 30–3; although I differ with some of his views here, our general conclusion is similar.

8 Jon Sumida, *In Defence of Naval Supremacy: Finance, Technology and British Naval Policy* (Boston: Unwin, 1989), Tables 3–14; Stevenson, *Armament and the Coming of War*, pp. 7–8.

9 USA Department of Defense, *National Defense Budget Estimates for FY 2004* (Mar. 2003), pp. 154–9, 172–7, 190–5 (on the WWW).

10 Ibid., pp. 148–53, 166–71, 184–9.

11 Ibid., pp. 160–5, 178–83, 196–201.

12 See chapter 2 in this volume.

13 Philip Cohen, 'A Terrifying Power,' *New Scientist*, Jan. 30, 1999, p. 10; Rachel Nowak, 'Disaster in the Making,' *op. cit.*, Jan. 13, 2001, pp. 4–5; Carina Dennis, 'The Bugs of War,' *Nature*, 17 May 2001, pp. 232–5.

14 Nadine Gurr and Benjamin Cole, *The New Face of Terrorism: Threats from Weapons of Mass Destruction* (London: Tauris, 2000), p. 43.

15 Spencer Hsu, 'Modest Gains against Ever-Present Bioterrorism Threat,' *Washington Post*, Aug. 3, 2008.

16 Anonymous scientist cited by Anne Applebaum, 'The Next Plague,' *The Washington Post*, Feb. 18, 2004.

17 David Ignatius, 'Portents of a Nuclear Al-Qaeda,' *The Washington Post*, Oct. 18, 2007.

5   Female participation in war
Bio-cultural interactions

From earliest times and throughout history, fighting has been associated with men. As we shall see, cross-cultural studies of male/female differences have found serious violence as the most distinctive sex-related behavioural difference there is. Is that a matter of education and social conventions, or are men naturally far more adapted to fighting than women are? This question, of course, has much contemporary relevance and is at the centre of a heated public debate throughout the West regarding women's equality in modern society: can and should women nowadays enlist in combat roles in the armed services? This chapter attempts to elucidate the nature of the bio-cultural interactions involved, whose complexity, and even existence, are all too often ignored in the debate. This may facilitate a realistic, cood-headed assessment of the possibilities and of future trends.

The first obvious and generally controversy-free fighting-related difference between men and women is that of physical strength. Men are considerably stronger than women, on average, of course, and all the following data is on average. To begin with, men are bigger than women. They are about 9 per cent taller and proportionately heavier. Even these facts do not tell the whole story, because in muscle and bone mass men's advantage is bigger still. Relative to body weight, men are more muscular and bony, with the main difference concentrated in the arms, chest, and shoulders. Fat comprises only 15 per cent of their body weight, compared to 27 per cent in women. As athletic results and repeated tests show, men's biggest physical advantage is in strength. While they are less flexible than women, only about 10 per cent faster, and have 4:3 advantage in aerobic capacity, they are doubly as strong as women (except for the legs, where the ratio is again 4:3 in favour of the men).[1] Since throughout human history fighting has been a trial of force, this sex difference has been crucial.

Anatomy is not everything, however. As mentioned, the quoted data is on average. It comprises in fact a wide range within each sex, and there is obviously some overlap between the scales of the two sexes. Some women are at least as strong as some men. Yet there is another sex difference to consider. Are men by nature mentally more aggressive than women, especially more predisposed to violence and, even more, to serious violence? Are the *minds* as well as the *bodies* of males and females different? This is a highly charged topic in the contemporary debate. *Tabula rasa* liberals and feminists during the 1960s and early 1970s believed that apart from obvious physical differences men and women were the same. All other differences were attributed to education and social conventions.[2] Over time, however, as more and more women entered the 'man's world' in the workplace and all other walks of social life, many later generation feminists have adopted a different position. They have come to feel that the 'man's world' was exactly that, very much structured to fit the needs, aims, and norms that were peculiarly masculine. They have felt that mere equality of access to male-structured domains was unsatisfying for women.

Gender attitudes to sex are one of the most interesting cases in point. One of the greatest achievements of the sexual revolution of the 1960s was that women in the developed world have earned the right to much the same freedom in sexual relations that men had always enjoyed. Soon, however, women discovered that they did not want to exercise that freedom in quite the same way as men. Thus, while latter-day feminists have continued to seek equality and opportunity, many of them now feel that these mean freedom to behave in greater harmony with women's own particular needs and aims, and, wherever necessary, change the world in that direction. Interestingly, it has now been feminists, not only male chauvinists, who have stressed women's qualities versus men's. Indeed, the more radical feminists have charged that it was peculiar male tendencies, such as over-competitiveness, emotional coldness, faulty communication, and aggressiveness that were responsible for many if not most of this world's ills, including war.[3]

These feminists may claim some support from the scientific research of human biology which earlier was all too often somehow regarded impatiently as irrelevant to the debate. The whole trend of recent scientific research has stressed sex differences in the mind as well as the body. Not only do gender attitudes to sex differ;[4] scientists have discovered many other differences. Repeated cognitive studies have revealed, on average, male advantage in spatial orientation, which might also explain the persistently recorded male advantage in mathematics, especially at the very highest levels. Women have recorded better in spatial attention to detail and spatial memory, verbal skills, and judging other people's moods and complex human situations – the famous 'feminine intuition'. These differences were long attributed solely to education and social expectations, but the great changes in social attitudes which have taken place in the last generation seem not to have altered them much. Indeed, one of the 'hardest' sciences of them all, brain research, has yielded significant sex differences. Cognitive studies, aided by brain scanning, have revealed that men and women in fact use different parts of their brains in coping with various cognitive tasks. Furthermore, whereas the right and left hemispheres of a man's brain are much more specialized, those of women operate in greater cooperation, and the *corpus callosum* connecting them is larger. Not only the bodies of women and men are structured somewhat differently but also that particular organ of their bodies, the brain, and, hence, their minds.

The architect of these different structures is our genes, and their agent is the sex hormones, particularly the famous male hormone, testosterone. Scientists have found that its presence is beginning to structure the male as different from the female right from the start, from the very beginning of the fetus' evolution in the womb (biologically, the original form is the female). Male and female differences in identity are already largely shaped at birth, and behavioural differences between the sexes are recorded very early, before social conditioning can play an effective role. Crudely put, baby girls are more interested in people, while baby boys in things. Later on, despite the great changes that have taken place in educational patterns and the efforts of conscientious parents, boys and girls show differences in play preferences, with the boys much more inclined to competitive, rough and tumble, and aggressive games and toys. Females also produce testosterone, only much less than males. In addition, some divergences from testosterone norms have occurred as a result of natural reasons (which produce identified medical syndromes) and owing to chemical influences caused for example by medication. It has been found that so-called 'tomboy' behaviour in girls correlated closely with higher levels of testosterone. On the other side, low testosterone levels in males result in unassertive and 'feminine' behaviour, while the highest levels of testosterone to which men are exposed during adolescence result in extra-aggressiveness.[5] Traditional, 'old fashioned' human insight embodied in such concepts as the Chinese *yin* and *yang* has been found to be not that far off the mark.

Perpetration of serious violence and crime are in fact the most distinctive sex-related behavioural difference there is, cross-culturally. Among the !Kung Bushmen of the Kalahari Desert, all of the 22 killings registered in 1963–1969 were committed by men. Of 34 cases of bodily assault, all but one were committed by men.[6] In the United States, males comprise 83 per cent of murderers, a similar share of those committing aggravated assault, 93 per cent of drunken drivers, and about the same percentage of armed robbers. Even though murder rates diverge widely in other parts of the world, the woman/man split remains roughly the same in favour of the men. Furthermore, even that sharp split does not tell the whole story.[7] The actual split is sharper still, for much of the serious female violence and murder comes in response to male violence or under male leadership. Thus, as a comprehensive survey reveals:

> Crime statistics from Australia, Botswana, Brazil, Canada, Denmark, England and Wales, Germany, Iceland, India, Kenya, Mexico, Nigeria, Scotland, Uganda, a dozen different locations in the United States, and Zaire, as well as from thirteenth- and fourteenth-century England and nineteenth-century America – from hunter-gatherer communities, tribal societies, and medieval and modern nation-states – all uncover the same fundamental pattern. In all these societies, with a single exception, the probability that the same-sex murder has been committed by a man, not a woman, ranges from 92 to 100 percent.[8]

This brings us to the nature of women's aggression and violence. Women can also be aggressive. However, their aggressiveness is much less channelled to physical violence than men's aggressiveness is, and even less to serious physical violence. Typically, women resort to serious violence in two cases, when the danger comes close to home: in desperate defence against an acute threat to themselves and their children; or to harm the 'other woman' in rivalry over a man. Furthermore, in comparison to men's violent aggression, that of women tends to be non-physical, indirect, and anonymous.[9]

What is the source of this most distinctive sex difference in serious violence? Again, the biological explanation is clear and was first elaborated by Darwin.[10] Both the bodies and minds of women and men have been subjected to somewhat different evolutionary pressures during the millions of years of human evolution. These pressures have been most different where sex specialization and diverging reproductive roles have been most involved. As scholars have pointed out, precisely because in humans both parents invest in child rearing, sex specialization/division of labour became more possible than in some other animal species, including our closest relatives, the chimpanzees. In evolutionary terms, women specialized in child bearing and rearing and in foraging close to the home base, whereas men specialized in long-distance hunting and in the struggle to acquire and defend women and children, specializations that required, among other things, force and ferocity. Indeed, the difference was more than occupational. Not only did men compete for women both inside and outside the group, often violently; in case of a threat to the children, the father, although also highly significant for the children's provision, was more expendable than the mother in this respect. For this reason as well, the men formed the group's main line of defence, while the women covered the children to the best of their abilities. Moreover, in the absence of agriculture and slavery, Palaeolithic men were of no use to the enemy. For them, the options were either running away or fighting to the finish. By contrast, women were themselves a resource in competition. They had better chances than the men did to survive the day by submitting, conforming, cooperating, and manipulating. Both the capabilities and evolutionary strategies of men and women, capabilities and strategies which were of course interconnected and mutually reinforcing, made men extremely more predisposed to fighting than women. Interestingly, modern day polls throughout the West consistently show significantly higher male than female support for policies which involve the military use of force.[11]

But do not environmental influences, most notably education and social norms, count at all? Do not genes always interact with culture? Obviously, environmental influences matter a great deal and are responsible for a wide diversity of cultural norms. However, contrary to the fashion in much of the gender studies, cultural norms are not infinitely flexible and wholly relative. As a rule, cultural norms play, and diverge, along a scale set by our inborn dispositions. Among hunter-gatherers, who represent the human way of life during more than 90 per cent of the human evolutionary history – what can be meaningfully referred to as the human 'evolutionary state of nature' – women's participation in warfare was extremely marginal. Even more than hunting, in which women also marginally engaged in a few societies, fighting was a male preserve and the most marked sex difference.[12] The 'culture of war' and the 'bond of brotherhood' within the warriors' group were famously cultivated among the men.[13] Again, it is interesting to note that modern day, and often feminist-motivated, studies register the following results: all-male groups are different in their overall attitudes (including to violence) from all-female groups; mixed groups are different still, while nevertheless maintaining male dominance.[14]

This does not mean that women had no role in warfare. In most cases they, too, had very high stakes in what the men were fighting for, or at the very least in their men themselves. Thus, women in so-called primitive warfare often accompanied the men to battle and took part in it as cheerers and providers of auxiliary services, such as the gathering and re-supply of used arrows and spears. As mentioned earlier, only in very rare cases did they actively participate in the fight, mainly by shooting arrows; and if the danger reached the inner ring of women and children, women also desperately tried to contribute to the defence. Indeed, bows and arrows, used from a far, lowered the obstacles to women's participation in fighting. The famous Amazons were, significantly, a myth, albeit, like many myths, not entirely devoid of some basis in reality. The Scythians and Sarmatians pastoralist horse archers of the Ukrainian steppe were described by the Classical Greek authors as the 'neighbours' of the Amazons. Some of the warrior graves excavated in the region were those of women, buried with full military gear. In one Scythian royal kurgan (mound) four out of 50 warrior graves belonged to females. In the supposed Sarmatian region, 20 per cent of the warrior graves excavated were those of women.[15]

Civilization created many new, 'artificial', conditions and relationships, making possible a far-reaching transformation in the human way of life. Nevertheless, throughout most of history, female participation in warfare barely changed at all from the patterns previously described, which had been evolutionarily shaped by physical, mental, and social constraints. Apart from desperate home defence, women's participation in warfare was limited to auxiliary services to the male warriors as camp followers and prostitutes. To be sure, women were excluded from many activities and occupations in historical societies. Still, they were absent from the warriors' ranks to an ever larger degree than from any other occupation in which they traditionally did not participate. But what about modern, industrialized, and especially advanced industrial societies? These have undergone tremendous, unprecedented, changes, which, among other things, greatly transformed women's place in society. How do these changes affect, and how can they affect, women's participation in combat roles in the armed services?

The bottom line is that they do, though overall perhaps not by a very wide margin. Physically, fighting with guns and explosives has already made a change. For example, in eighteenth and nineteenth centuries Dahomey, the king's army included an elite bodyguard unit of women, which grew in number from hundreds to thousands. The women, armed with guns, as well as with bows and arrows, machetes and clubs, were reputedly ferocious warriors.[16] From the late nineteenth century, women began to participate actively in many revolutionary and guerrilla forces, which combined informal social structures and radical ideologies. Their participation in combat roles in the Soviet and Yugoslav armed forces during the Second World War and on the communist side in Vietnam is well known. However, even in these often-cited cases, where a radical social ideology prevailed, where the home country was invaded and women were anyhow at grave risk, and where an acute shortage of manpower existed, women's role in warfare was still limited. Most women took men's place in the factories and fields, or performed auxiliary services within the armed forces. Those who actually participated in combat roles amounted to no more than 8–12 per cent of the combat troops, not far from their estimated share in the famous Dahomey army and in those very few tribal societies who had allowed women to participate in battle, including the Scythian and Sarmatian 'Amazons'. Furthermore, in Soviet Russia, Yugoslavia, Vietnam, and other revolutionary countries alike, the women were excluded from combat roles once the war was over.[17] Typically, a study of women's participation in the Nicaraguan revolution, boldly entitled 'No Going Back', ends with the sobering fact that after the revolution was over women *did* go back, at least from the armed forces.[18]

Why is this so, and how likely is this situation to persist in advanced industrial societies? After all, the modern mechanical and electronic battlefield has created numerous tasks that involve little if any physical force. Fighting is done with firepower, and the movement of people and loads is largely mechanical. Many women can drive or fire a tank as well as many men, or for that matter command the tank, a tank battalion, or a tank army. Some women are even strong enough to be able to serve in ordinary infantry units, which still rely heavily on physical force. However, Hollywood's *G.I. Jane* notwithstanding, women are rarely likely to be strong enough for elite infantry and crack units; no more in fact than they are likely to compete successfully in any serious men's football league, let alone boxing or weightlifting. Women flew as combat pilots in the Soviet air force during the Second World War. But how many of them can successfully compete for similar capacities in the much more competitive air forces of modern advanced powers has still to be experienced. In any case, this leaves many active combat activities that women can perform.

The mental sex differences in respect to warfare have similarly narrowed but not closed. Since much of today's fighting activity is done from afar and with little physical contact, it involves much less of the aggressive and violent attitude traditionally associated with men. Even if not wholly a matter of pushing buttons, modern fighting more than before bears the character of an occupation that requires more cool-headed professionalism and organizational discipline than aggressive predisposition. There can be little doubt that many women could cope successfully with the mental task if they so wished. But would they so wish? The indications are that the number of those who would wish it is far smaller than that of men. Even if the physical aspect posed no problem, far fewer women than men are *inclined* to combat activity and combat careers. The reasons for this motivational difference again go back to fundamental sex-related predispositions. On average, men are more attracted to this type of competitive, high-risk, violent, machine-related activity. In the same way that the introduction of effective contraceptives, although greatly affecting women's sexual attitude, has not closed the gap between the sexual behaviour of men and women, far-reaching changes in social and family patterns do not wholly eradicate sex-related occupational preferences.

Throughout history women's overburden with child bearing and rearing was one of the factors that precluded their active participation in warfare. Indeed, significantly, the famous Dahomey women warriors unit was only possible because its members, officially married to the king, were forced to celibacy on penalty of death. The force may have evolved from the harem guard, to which no man was allowed access. Furthermore, the women may have customarily undergone excision at childhood.[19] Even though women in today's developed world only give birth to fewer than two children, on average, and household duties are far lighter than before and are more equally divided between the sexes, the woman's share in raising the children still tends to be larger. More than men, women tend to recoil from a highly risky career that involves long periods of absence from the husband and children. This sort of preference has long been attributed to lingering cultural inequalities in the way society is structured. While these inequalities were indeed acute and still exist, it would now seem that their inborn element has been too easily overlooked. Even if the greatest equality of access to the educational and labour markets were achieved, the sex differences are such that the inclinations of men and women would be, on average, different in some important respects. Even in Scandinavia, where nearly 80 per cent of women are in the workforce, fewer than 10 per cent of the women work in occupations where the sex balance is roughly equal. Half of all workers are in jobs where their own sex accounts for 90 per cent of employees.[20] The choice of a combat career is a field in which the sex difference is particularly marked.

The Netherlands are a case in point, having the most egalitarian legislation and policy in the developed world. From the late 1970s the Dutch authorities gave women equal access to all military jobs and have acted intensively to encourage them to exercise this freedom of opportunity. For all that, as the feminist authors of a study on the subject have written with dismay:

> The interest of women in the army seemed to diminish more than to increase. . . . The physical requirements remained a problem and so did the acceptance of women by their male colleagues. . . . The demands for combat jobs in the infantry, cavalry, artillery and the Royal Engineers are too high to be met by most women.

Female participation in the army, especially in combat roles, remained in the low percentage points. In Norway as well, another country with highly egalitarian legislation and policy, the picture is very similar, partly, although not solely, because of women's own lack of interest.[21]

But what about those women who do desire a combat role and a combat career? In the labour market as well, many occupations are unevenly divided between the sexes, but equality of access on merit has nevertheless been secured in the developed countries to any member of either sex who chooses any particular occupation. Are there any special arguments that might warrant an exceptional status to the occupation of fighting? More complex family arrangements, mentioned by reluctant armed services, have already been discussed. These may be overcome by a combination of female and military compromises. The prospect of possible captivity is a major consideration. Gender differences in relation to sex are such that women are far more exposed than men to sexual abuse, especially when out of the protection of the law and of orderly society. This, too, however, is a risk that society might choose to leave to individual female choice. Finally, expressing wide and deeply held concerns: can men and women be close together for long periods of service in intimate combat groups without being distracted by sexual attraction that would seriously disrupt their combat effectiveness? Does not the famous 'male bonding' in the combat group depend on the absence of women? Is not the 'culture of war' itself, those traditional qualities of warrior masculinity, best inculcated in an exclusively man's world? Indeed, at this point some feminists form an awkward alliance with male sceptics, arguing that experience shows that participation in combat units make women forfeit their own true nature and adopt male-type thinking and behaviour.

We lack sufficient experience to judge how significantly the dynamics created in modern mixed-sex fighting units would affect their combat effectiveness. In principle, fighting units need not, of course, necessarily be mixed for women to participate in them. Separate units for men and women are also possible. In summary, the heated rhetoric and polarized positions of the public debate may conceal a more complex reality. It would probably not be a wild speculation to suggest that the forces that have opened the labour market for women are too irresistible for the armed services to withstand. Women will be integrated in larger numbers, surely in auxiliary and combat-support roles, and even in combat roles. On the other hand, expectations embodied in titles such as Jeanne Holm, *Women in the Military: An Unfinished Revolution* (1982) appear to be misplaced. Women's participation will probably remain marginal compared to that of men.[22] The evolution-shaped physical, mental, and social factors which have made fighting the most polarized sex-related activity are unlikely to disappear.

Notes

1 Studies are summarized in Mary Anne Baker (ed.), *Sex Differences in Human Performance* (Chichester: Wiley, 1987), esp. pp. 109–10, 117, 127, 136–7, 180; also Donald Symons, *The Evolution of Human Sexuality* (New York: Oxford UP, 1979), p. 142; Marvin Harris, *Our Kind* (New York: HarperCollins, 1990), pp. 277–81.
2 In the psychological research this attitude dominated Sleanor Maccoby (ed.), *The Development of Sex Differences* (Stanford: Stanford UP, 1966).
3 Support for this view with respect to war from a sociobiological point of view is offered in Richard Wrangham and Dale Peterson, *Demonic Males: Apes and the Origins of Human Violence* (London: Bloomsbury, 1997); for works in this vein see p. 284, n. 53; Berenice Carroll and Barbara W. Hall, 'Feminist Perspectives on Women and the Use of Force,' in R. Howes and M. Stevenson (eds.), *Women and the Use of Military Force* (Boulder and London: Lynne Rienner, 1993); Annette Fuentes, 'Women Warriors? Equality, Yes – Militarism, No,' reprinted in E.A. Blacksmith (ed.), *Women in the Military* (New York: Wilson, 1992), pp. 34–40; Helen Vozenilek, 'Women Should Not Support America's Military Goals,' and Mary Hunt, 'Medals on Our Blouses?', both reprinted in C. Wekesser and M. Polesetsky (eds.), *Women in the Military* (San Diego: Greenhaven Press, 1991), pp. 50–2, 96–100 respectively; also the works cited in Wendy Chapkins, 'Sexuality and Militarism,' in Eva Isaksson (ed.), *Women and the*

Military Service (New York: St. Martin's Press, 1988), p. 106; Barbara Ehrenreich, *Blood Rites: Origins and History of the Passions of War* (New York: Metropolitan, 1997), pp. 125–31.

4 See esp. Symons, *Human Sexuality*; Matt Ridley, *The Red Queen: Sex and the Evolution of Human Nature* (New York: Macmillan, 1994).

5 Studies are summarized in R.M. Rose et al., 'Androgens and Aggression: A Review of Recent Findings in Primates,' in Ralph Holloway (ed.), *Primate Aggression, Territoriality, and Xenophobia* (New York: Academic Press, 1974), pp. 276–304; E.E. Maccoby and C.N. Jacklin, *The Psychology of Sex Differences* (Palo Alto, CA: Stanford UP, 1974); Luigi Valzelli, *Psychobiology of Aggression and Violence* (New York: Raven Press, 1981), pp. 116–21; Anne Moir and David Jessel, *Brain Sex: The Real Difference between Men and Women* (New York: Lyle Stuart, 1991); M. Daly and M. Wilson, *Sex Evolution and Behaviour* (Belmont, CA: Wadsworth, 1983); Felicity Huntingford and Angela Turner, *Animal Conflict* (London: Chapman, 1987), pp. 95–128, 339–41; J. Herbert, 'The Physiology of Aggression', and Marshall Segal, 'Cultural Factored Biology and Human Aggression,' in J. Groebel and R.A. Hinde, *Aggression and War* (Cambridge: Cambridge UP, 1989), pp. 58–71, 173–85; Ridley, *The Red Queen*, pp. 247–63; James Wilson, *The Moral Sense* (New York: Free Press, 1993), pp. 165–90.

6 Richard Lee, 'Politics, Sexual and Non-Sexual, in Egalitarian Society,' in Richard Lee and E. Leacock (eds.), *Politics and History in Band Societies* (New York: Cambridge UP, 1982), p. 44.

7 Ridley, *The Red Queen*, p. 252; Wrangham and Peterson, *Demonic Males*, pp. 113, 115; Segal, 'Cultural Factored Biology and Human Aggression,' pp. 177–8; David Jones, *Women Warriors: A History* (Washington, DC: Brassey's, 1997), p. 4.

8 Wrangham and Peterson, *Demonic Males*, p. 115, based on Martin Daly and Margo Wilson, *Homicide* (New York: Aldin, 1988), pp. 147–8. Denmark is the exception with 85 percent for the males, but omitting infanticide the figure rises to 100 percent.

9 Huntingford and Turner, *Animal Conflict*, pp. 332–3; K. Bjorkqvist and P. Niemela (eds.), *Of Mice and Women: Aspects of Female Aggression* (Orlando, FL: Academic Press, 1992); Kirsti Lagerspetz and Kaj Byorkqvist, 'Indirect Aggression in Boys and Girls,' in L.R. Huesmann (ed.), *Aggressive Behavior: Current Perspectives* (New York: Plenum Press, 1994), pp. 131–50.

10 See especially Symons, *Human Sexuality*; Ridley, *The Red Queen*; Frans de Waal, *Good Natured: The Origins of Right and Wrong in Humans and Other Animals* (Cambridge, MA: Harvard UP), pp. 117–25. Darwin's (sharp) distinction in 'Sexual Selection in Relation to Man' may be too Victorian for our taste: *The Descent of Man*, Ch. xix, in *The Origin of the Species and the Descent of Man* (New York: The Modern Library, n.d.), pp. 867–73.

11 Nancy Gallagher, 'The Gender Gap in Popular Attitude towards the Use of Force,' in Howes and Stevenson, *Women and the Use of Military Force*, pp. 23–37.

12 For a few known cases of women's participation in warfare in tribal societies see Maurice Davie, *The Evolution of War* (New Haven: Yale UP, 1929), pp. 30–6; David Adams, 'Why There Are So Few Women Warriors,' *Behaviour Science Research*, 18 (1983), pp. 196–212, whose explanation for this rarity is otherwise naïvely simplistic; Walter Goldschmidt, 'Inducement to Military Participation in Tribal Societies,' in R. Rubinstein and M. Foster (eds.), *The Social Dynamics of Peace and Conflict* (Boulder, CO: Westview, 1988), p. 57.

13 Lionel Tiger, *Men in Groups* (New York: Random House, 1969), pp. 80–92.

14 Jill Bystydzienski, 'Women in Groups and Organizations: Implications for the Use of Force,' in Howes and Stevenson, *Women and the Use of Military Force*, pp. 39–52.

15 Timothy Taylor, 'Thracians, Scythians, and Dacians, 800 BC–AD 300,' in Barry Cunliffe (ed.), *The Oxford Illustrated Prehistory of Europe* (Oxford: Oxford UP, 1994), pp. 395–7.

16 There is now a book on this subject: Stanley Alpern, *Amazons of Black Sparta: The Warriors of Dahomey* (New York: New York UP, 1998).

17 See Nancy Loring Goldman (ed.), *Female Soldiers – Combatants or Noncombatants?* (Westport, CT: Greenwood Press, 1982), esp. pp. 5, 73, 90, 99; scholarly inferior and uneven is Isaksson (ed.), *Women and the Military*, but see esp. pp. 52–9, 171–7, 204–8; Jones's anecdotal and scholarly shaky *Women Warriors* throughout corroborates the traditional picture of the limits of women's participation in warfare which he sets out to refute.

18 Barbara Seitz, Linda Lobao, and Ellen Treadway, 'No Going Back: Women's Participation in the Nicaraguan Revolution and in Postrevolutionary Movements,' in Howes and Stevenson, *Women and the Use of Military Force*, pp. 167–83.

19 Alpern, *Amazons of Black Sparta*.

20 Matt Ridley, *The Origins of Virtue: Human Instincts and the Evolution of Cooperation* (New York: Viking, 1996), p. 93.

21 Annemiek Bolscher and Ine Megens, 'The Netherlands,' in Isakson, *Women and the Military*, pp. 359–69; Ellen Elster, 'Norway,' *op. cit.*, pp. 371–3.

22 Interestingly, Mady W. Segal, 'Women's Military Roles Cross-Nationally: Past, Present, and Future,' *Gender & Society*, 9 (1995), pp. 757–75, reaches similar conclusions. Although committed to the idea that gender roles are socially constructed and leaving biology out of her model of factors influencing women's military participation, biology (inevitably as a constant) nevertheless creeps in in her discussion and surely must account for her results.

6 Is democracy genocidal?

The liberal democracies' colonial record includes a particularly problematic element: the charge that in both the United States and Australia, democracies exterminated the native populations. This is the source of a profound sense of guilt in the two countries, reinforcing pervasive doubts about whether liberal democratic societies really behave better than others. Leading historical sociologist and left-wing ideologist Michael Mann develops the charge of genocide in his book *The Dark Side of Democracy: Explaining Ethnic Cleansing* (2004). Mann regards it as a major flaw in the democratic peace proposition. But this charge is fundamentally invalid, as are the conclusions derived from it.

Mann's main thesis is that murderous ethnic cleansing, which in extreme forms can become genocidal, is predominantly a modern phenomenon. It is the 'dark side of democracy', when the rule of the people and the ethnicity of the people are 'confused'. In premodern times, class prevailed over ethnicity in the eyes of conquerors and social elites, which sought to subjugate and exploit people irrespective of their ethnic identity, rather than get rid of them. However, the advent of modernity, popular sovereignty, and universal citizenship occasionally spurred violence between rival ethnic groups that claimed the same territory. In extreme cases, their struggles escalated into murderous ethnic cleansing and even genocide.

We begin our scrutiny of this thesis with the premodern world. Was murderous ethnic cleansing rare then? As revealed by recent scholarship, prehistoric conflict between small human groups often resulted in massacre and even extermination. A singular example is the spread of our species, *homo sapiens*, out of Africa, from about 60,000 years ago onward. In that process, all the archaic humans who inhabited the Old World were gradually extinguished, everywhere displaced and replaced, victims of what was probably our species' superior skills in subsisting, reproducing, and fighting. Incidentally, there is no need to shift the blame for murderous mass killing from our civilization to our species, a popular holdover from the 1960s. As viewers of television nature documentaries witness, high rates of intra-killing within animal species are the norm in nature.

Another example may be the displacement and replacement of the earliest inhabitants of what is present-day Japan by the population now known as Japanese. The Japanese apparently arrived from Korea in around 300 BC, bringing wet rice agriculture with them. They gradually pushed the earlier, Jamon, inhabitants up the archipelago through a combination of numbers, dense agricultural settlement, and warfare. Today only about 150,000 of the remnant of the Jamon, the Ainu, live in Hokkaido and other northern islands. And this is merely one example of many from the shadowy light of prehistory.

We now move from prehistory to history, when states, stratified societies, and elite rule dominated. This is an era that is far more central to Mann's argument. He notes the massive killing, massacre, razing of cities, and mass uprooting of populations that were very much the stuff of history. He mentions some well-known examples from the wars of Assyria, Greece, Rome, Carthage, the Anglo-Saxon invasion of Britain, and the exploits of nomadic hordes such as the Mongols. As he correctly states, 'Warfare occasionally strayed into ethnocide.' All the same, Mann insists that since economic exploitation by a ruling and often ethnically foreign elite was the rationale of premodern society, such brutality should be regarded as marginal. According to this logic, the conquering and ruling elites resorted to vicious means to suppress and/or set an example for others as quickly and efficiently as possible, with the aim of resuming exploitation with the least possible disturbance.

But this attractive model is misleading. For one thing, enslavement was always one of the main options of exploitation, and not only in the figurative sense of subjugating people in and on their land, but also in the ordinary sense (which Mann barely mentions) of taking them as captives and selling them far away from home. This was standard practice in military operations, often constituting the main booty. Occasionally, conquering armies emptied cities and bled provinces white. Indeed, we sometimes hear of the price of slaves falling in the markets on account of the massive supply. Africa, of course, furnishes an especially gruesome example: tens of millions of people were carried north through the Sahara and out of East Africa by Arab and Muslim slave traders well before Europeans began the trade from West Africa. Both Arabs and Europeans collaborated with native polities that provided the slaves through raids and wars against their neighbours. Thus ethnic cleansing was economically motivated well before modernity.

Furthermore, the model that Mann seems to have in mind is that of imperial wars. However, routine warlike activity in history was much more mundane, taking place between neighbouring small-scale polities and driven by a variety of motives that most often included capturing border land. While the victors grabbed and settled on the land, the indigenous population typically fled rather than waiting to be looted, enslaved, driven away, or killed. In most cases, such small polities did not have the capability to rule over one another, and on the rare occasions that one of them succeeded in conquering a long-time neighbour and rival, it more characteristically razed it to the ground. If 'macro-ethnic cleansing' was rare in premodern times, it is because macro-ethnicities and macro-states themselves were rarer. Murderous 'micro-ethnic cleansing' took place all the time.

So ethnic cleansing, involving widespread massacres, mass destruction of settlements, mass enslavement, and masses of refugees driven out by war, were commonplace in premodern times – probably more so than in modern times. All the same, Mann may still have a point in claiming that modern ethnic cleansing is somehow different, possessing a special quality that distinguishes it from earlier examples. As he sees it, democracy is the new development that begot ethnic cleansing. His point would carry some validity, but for the strange meanings he attaches to the concept of democracy.

It is true that spreading notions of popular sovereignty and equal universal citizenship made the peaceful existence of ethnically foreign people within a state more problematic, and prompted deep tensions and violence. The growth of popular sovereignty, the institutionalization of universal citizenship, the creation of mass society, and deepening popular mobilization are sometimes reasonably labelled democratization. Some scholars of nationalism point out the connection of all these to the rise of nationalism. There was no confusion here between the political and ethnic meanings of the concept of the people, as Mann would have it, criticizing President Wilson, for example, for committing such a category mistake. Rather, as people's involvement in the public sphere increased, they exhibited a strong preference for political self-identification and self-determination in concert with their ethnic kin. That preference has profoundly and often violently affected both domestic and international politics during the past two centuries. In this sense, then, democratization and nationalism have gone hand in hand (see my *Nations: The Longe History and Deep Roots of Political Ethnicity and Nationalism*, 2013). However, whereas popular sovereignty, universal citizenship, mass society, and popular mobilization can be labelled democratization, they cannot be equated with democracy, a far more restricted phenomenon, without stretching the latter term to the breaking point. The great majority of state societies that underwent these processes did not become democratic, unless the term is indiscriminately applied to a whole range of repressive and often atrocious authoritarian and totalitarian regimes.

Mann examines a series of ethnic cleansing and genocide cases during the nineteenth and twentieth centuries that he surprisingly attributes to democracy: ethnic expulsions and massacres in the nondemocratic Balkans before and after the First World War; Czarist Russia's treatment of Muslims in the conquered Caucasus; the genocide of the Armenians in the Ottoman empire under the authoritarian-nationalist Young Turks; the deportation of Germans from post-World War II Eastern Europe; and the events in the democratizing former Yugoslavia and Rwanda during the 1990s. Even more surprisingly, he also includes in the list of democratic ethnic cleansing the Nazi genocide of the Jews and their actual and planned treatment of the Slavs, as well as communist cleansings, mass killings, and genocides under Stalin in the Soviet Union, Mao in China, and Pol Pot in Cambodia. In Mann's terminology there are 'liberal democracies' and 'ethnic democracies', with Nazi Germany belonging to the latter type. The accepted distinction is actually between liberal or civic nationalism and ethnic nationalism. Although this distinction has its problems, it is less confusing and it more clearly indicates the real culprit: militant nationalism.

Of course, the totalitarian socialist countries such as Russia, China, and Cambodia can hardly be classified as ethnic democracies, so Mann further overstretches his thesis by claiming that Stalin, Mao, and Pol Pot in fact carried out their class persecution and elimination in the name of a class identified as 'The People', and thus fall under the category of democratic (and ethnic?) cleansing. This explanation recalls the old designations of these totalitarian socialist regimes as 'people's republics' and 'democratic republics'.

Why would an able sociologist like Mann tie himself in such strange knots? It is because he believes that not only 'ethnic democracies' but also liberal democracies strayed into ethnic cleansing in colonial-frontier settings. He alleges that genocidal democracies were responsible for 'the most successful cleansing the world may have ever seen': the United States targeted the Native Americans and Australia the Aboriginals. Both indigenous groups were displaced from their lands and experienced, respectively, an estimated 90 per cent and 80 per cent drop in numbers during the nineteenth and early twentieth centuries. Mann concedes that well-established democracies do not carry out murderous ethnic cleansing, but claims that in earlier, more formative stages of their development they did.

To be sure, both Native Americans and Australian Aboriginals were victims of massive expropriation, and their violent resistance was suppressed by the more effective violence of governments and local white settlers. The land on which the natives lived was progressively taken from them, devastating their livelihood and way of life. The United States carried out some large-scale and deadly deportations, such as that of the Cherokees, and local militias, most notably in California, engaged in indiscriminate killings. Small-scale violence and killings were pervasive. Native women were occasionally taken away and raped by white men. The natives' lives, well-being, and culture were horribly damaged. Most of the white population saw the natives as a menace to be defeated, as did the authorities, more so in the independent republic of the United States than in colonial Australia. The greatest presidents of democratic America – George Washington, Thomas Jefferson, Andrew Jackson, Abraham Lincoln, Theodore Roosevelt – prophesized the natives' ultimate doom, which they believed to be much-deserved in light of Native Americans' acts of savage violence. All this is undeniably true. Hence people tend to assume the worst. Nonetheless, the demographic catastrophe that befell the US Native Americans and the Australian Aboriginals was not caused by such means. The culprit was very different, and it struck indiscriminate of intent and regime type.

The source of the natives' demographic calamity in both the Americas and Australia was their lack of immunity to Old World diseases, including smallpox, measles, influenza, typhus, and tuberculosis. Most of these diseases had passed from domesticated livestock to humans during the Neolithic agricultural revolution and spread through the population of the Old World, which in time had developed resistance to them. The sudden joining of the Old and New Worlds spelled disaster for the natives of the latter, as the diseases migrated across the oceans with the European newcomers. Mann clearly presents the consequences of the ensuing biological holocaust, affirming the now prevailing view that it caused the death of some 90 per cent of the pre-contact native population throughout the Americas. He well recognizes that this horrific death toll was unintentional, rightly regarding the well-known sporadic cases in which contaminated blankets and cloths were deliberately given to the natives to precipitate their demise as inconsequential to the general, practically unstoppable, trend. All the same, while writing at great length about everything else, he devotes only a couple of pages to this factor, practically ignoring the implications of that devastating development for his argument.

But how can we determine precisely what part of the demise of these native populations was due to epidemic diseases and what part to human brutality? In history and the social sciences, where events cannot be replayed, the only means of answering such questions is through controlled comparisons. Mann argues correctly that in both the United States and Australia, where farmer settlement was the norm, the farmers had little use for the natives and were particularly interested in seeing them gone. For this reason, he believes, the frontier democracy practised by settlers was the most atrocious toward the natives. But what, then, can we learn from other places, where the relevant circumstances were different?

Let us begin with colonial Spanish America. Although from time to time Mann likes to chide 'our civilization' and 'Europeans', by association, as if they meant democracy, Spanish America was not democratic, of course. Furthermore, the Spanish conquistadores in the early sixteenth century were quintessentially premodern in Mann's sense: they exhibited great ruthlessness during their conquests and suppressed and abused the natives, but they wanted them to live so that they could be exploited. They needed the natives to work for them in mines and on landed estates or plantations – the common agricultural possession in Spanish America. However, the natives of the Caribbean, where the Spanish first landed and established their rule, died of European diseases at such a fast rate that they were wiped out altogether. The Spaniards saw no alternative but to import African slaves, who despite great mortality rates during ocean crossings and horrible abuse over centuries thereafter, survived in great numbers in the Americas because of their natural resistance to Old World epidemic diseases.

Much the same applies to the rest of Spanish America, where European diseases were a major factor in the destruction of the Aztec and Inca empires, killing an estimated half of each population, including emperors of both empires at the time of conquest. The common estimate now is that within a century after the conquest, the native population fell by about 90 per cent and then slowly recovered in the following centuries, after the people gradually developed resistance to the new diseases. The natives of Amazonia, isolated until recently, have been much abused over the past generation by entrepreneurs and their workforces penetrating the rain forests to carry out large-scale projects of economic development. Above all, however, their lives are threatened by contact with European diseases, whose deadly effects are today preventable by large-scale immunization.

We now turn to North America. A dense agricultural-urban civilization existed in the Mississippi River Valley, but a Spanish probing expedition led by Hernando De Soto observed in 1540 that its towns were deserted. This native civilization is believed to have succumbed to European diseases that spread from Mexico even before its contact with the Europeans, let alone with white settlers. When the English and French arrived at the same area in the following centuries, this civilization was already long gone. The same terrible process took place throughout

North America, as contact and disease gradually spread across the continent. Sparse native populations, rather than falling victim to genocide, were afflicted each in turn by disease brought in by the settlers. Most native populations reached their lowest point some one hundred years after contact. The natives of California suffered a disastrous decline, to an estimated half of their pre-contact population, under Spanish and Mexican rule and the mission and estate system. Their numbers continued to fall with the advent of American rule – again only minimally due to direct killings, cruel and indiscriminate as they were. The natives of Pennsylvania and New Jersey were sheltered from harassment by the Quaker communities alongside which they lived; this did not, however, prevent their virtual elimination by disease.

One last control case is that of French Algeria. The atrocious treatment of black Africans in America did not cause blacks to decline demographically – quite the opposite. Nor did colonial abuse in black Africa itself in the nineteenth and twentieth centuries have that effect, except for a few cases of direct genocide, such as those perpetrated by Imperial Germany (see chapter 7 in this volume). In North Africa, too, French occupation and settlement of Algeria after 1830 involved ferocious 'pacification' of native resistance. Nevertheless, during the very same period in which the native North American population was plummeting, Algeria's population actually surged, from some 2.5 million in 1800 to about 6 million in 1920.

Thus an examination of Mann's independent variables – democracy and the 'frontier democracy' of farmer settlers – in diverse cases across historical experience belies his contention about democracies' track record. This is not to say that the natives of America and Australia were not treated harshly and often atrociously. The advance of dense agricultural settlement forced them from their land and disrupted their livelihoods, their way of life, and their culture. They experienced blatant discrimination and were subject to deadly deportations and killings. These purposeful acts contributed to their disasters, but did not – could not – cause them. More generally, it is true that popular sovereignty, universal citizenship, mass popular mobilization, and mass society – which can be labelled democratization – made nationalism more manifest and, consequently, added a new dimension to ethnic tensions and cleansing. It is also true that liberal democracies' standards for the treatment of ethnic minorities in colonial settings were very different from those they practised at home, and were sometimes harsh. Liberal democratic standards rose dramatically during the nineteenth and twentieth centuries. Still, it takes ideological zealotry, which leads to fundamental errors in argument and method, to label ethnic cleansing as the dark side of democracy.

7   Why counterinsurgency fails[1]

Down to the present Iraqi and Afghan entanglements, insurgency warfare has earned a reputation for near invincibility, driving Britain and France out of their former colonial empires during the twentieth century and frustrating American military interventions even where the asymmetry in regular force capability has been the starkest. Why have mighty powers that proved capable of crushing the strongest of opponents like Imperial and Nazi Germany and Imperial Japan failed to defeat the humblest of military rivals in some of the world's poorest and weakest regions?

It is argued here that, rather than being universal, this difficulty has overwhelmingly been the lot of liberal democratic powers – and has been encountered precisely because they are liberal and democratic.[2] While it may seem obvious, this proposition is still far from being generally recognized by scholars and in the public discussion at large.[3] The crushing of an insurgency necessitates ruthless pressure on the civilian population, which liberal democracies have found increasingly unpalatable in cases other than total war for survival. Premodern powers rarely had a problem with such measures; nor have modern authoritarian and totalitarian powers been reluctant to use them. And overall they have proved quite successful in suppression. Indeed, suppression was the *sine qua non* of their imperial rule. Weakness in counterinsurgency is peculiar to the liberal democracies' conflict behaviour.

The secret of suppression and the liberal state

Throughout history, imperial pacification rested on the overt threat and actual application of ruthless violence to crush resistance in subject societies. Where the people offered insurgents support and sympathy, they were exposed to sweeping reprisals by the ruling power, including killing, looting, burning, and enslavement. The *ultima ratio* of imperial control was the threat of genocide. All empires worked this way, including democratic and republican ones, such as ancient Athens and Rome. They could only work this way.

During the Peloponnesian War, the Athenians reneged on their earlier decision to kill all the men and enslave the women and children in conquered Mytilene, which had defected from their empire (428 BC). Instead, they opted for the execution of 'only' some 1,000 men held responsible for the rebellion, which, given the size of the polis, still amounted to a very large part of its male population. The Athenian leader Cleon was nonetheless dismayed by this show of leniency,[4] but his expressed fear that democracies were incapable of ruling others turned out to be unsubstantiated (or at least premature), as the Athenians' famous dialogue with, and ultimate annihilation of, the people of Melos (416 BC) chillingly demonstrates. Republican Rome's record in dealing with insurgents was even more merciless.

However, the conduct that had sustained empires throughout history became increasingly unacceptable in the emergent liberal societies of the modern era. As with other elements of the liberal/democratic peace, this was a gradual and uneven process. Some mellowing of practices towards civilian populations was already discernable in Western Europe during the Age of Enlightenment. Yet in Europe's more backward areas, and with respect to non-whites, practices remained very much as before. Despite sporadic British atrocities, the rebelling American colonists benefited from this mellowing, intermixed as it was with the Royalist desire not to further alienate the colonial population.[5] But the people of Ireland – that 'Africa in Europe' for the British – rebelling in 1798, still bore the brunt of the British suppression that had broken their backs in blood and fire in earlier uprisings in previous centuries, and had crushed forever the rebellious 'savage' Scottish Highlanders after the Battle of Culloden (1746). The Americans' treatment of the Native Americans during the nineteenth century (though the great majority of them fell victim to European diseases; see the previous chapter in this volume) was similarly legitimized by their perception of them as savages. The French 'pacification' of Algeria and Indochina during the nineteenth century still relied on the old methods. Notably, however, Marshal Bugeaud's methods in Algeria were denounced by a delegation of the Chamber of Deputies, headed by Alexis de Tocqueville, which recommended the adoption of a 'Continental standard of conduct.'[6] The bloody Indian Mutiny (1857) was the last that the British Empire suppressed atrociously, in the old ways, though in this case, too, it should be noted that the troops' retributions lacked official sanction.

The establishment of Britain's Liberal Party in 1859 marks a symbolic turning point in British attitudes. That this had a significant impact on various aspects of British international conduct can be seen in the British attitude towards the Russian-supported rebellion of the Bulgarians against their Ottoman masters (1875–1878). It had been Britain's policy to back the Ottoman Empire against any Russian advance towards the Mediterranean, but the British public's outrage at the Turkish atrocities in suppressing the Bulgarian rebellion – the mass killings, torture, and rape – fueled by journalistic reports and by William Gladstone's missionary agitation, tied the British government's hands. The Ottoman Empire was defeated by Russia and forced to give up the rebellious province. 'Foreigners don't know what to make of the movement; and I am not surprised,' Foreign Secretary Lord Derby told Prime Minister Benjamin Disraeli. A German observer noted that it would be almost inconceivable in any Continental country.[7] No longer was British policy conducted on purely 'realist' considerations of power. In mid- and late-Victorian Britain, human rights became inseparable from the public debate on foreign policy.

Soon enough, the same attitude arose with regard to Britain's own empire, first in its white parts but later everywhere. The same Gladstone, as a Liberal prime minister, opened the process that would lead the Irish within a generation to an independent state, after the liberal recipe for self-determination within Britain failed to satisfy them. How did a country that had been under the British heel for centuries suddenly succeed in seceding? The rise of modern nationalism in Ireland cannot be the only explanation, because national movements have been successfully curbed and crushed by ruthless nonliberal powers. It was only when liberals could no longer resist the demand for self-determination, as well as finding the old methods of forceful suppression repugnant and unacceptable, that Ireland was able to gain independence. Needless to say, the process was anything but smooth. The Liberal Party was split over the Irish question and lost power for two decades. The Easter Uprising in Dublin in 1916 was put down by robust force, and a full-scale insurgency took place between 1919 and 1921, when Britain decided to pull out. Although British counterinsurgency tactics had proved quite effective, they could never completely quell the rebellion, given the restriction on ruthlessness towards civilians under which British forces operated.

Nor was Ireland an isolated case for Britain. In the Boer War in South Africa (1899–1902), Britain initially suffered humiliating defeats in regular fighting at the hand of the forces of the Free Orange and Transvaal Republics. When a half-million British troops were dispatched to South Africa, the course of the war was reversed and regular Boer resistance was crushed, only to give way to widespread irregular resistance. Unable to subdue that resistance, the British resorted to draconian measures, rounding up the Boer civilian population into concentration camps. Some 30,000 people perished in the camps from various illnesses. And yet Britain was able to declare victory only by offering the Boers the most generous of peace terms, which within a few years effectively surrendered to them government powers over all of the South African Federation. South Africa and Ireland were signs of things to come in liberal democracies' counterinsurgency wars.

Alternative explanations and queries

Since the American humiliation in Vietnam, a variety of explanations have been advanced to account for the puzzle of the weak defeating the strong. For one, it has been suggested that the problem lies in the number of troops and resources the strong country can spare for a particular local war, given its overall commitments and the difficulties of power projection to faraway theatres.[8] Most recently, the American problems in pacifying Iraq after its occupation in 2003 have been widely attributed to the commitment of too small a force (as if the deepening and failed American commitment in Vietnam had never taken place). Admittedly, security conditions have considerably improved following the American troop reinforcement during the 'surge'; but without a far-reaching indigenous change, most notably in the attitude of Sunni tribal leaders, the surge would have yielded little fruit and an American victory would have remained elusive. Massive force commitment and considerable military success ultimately failed to keep France in Algeria and Israel in densely-populated Palestinian territories (or in Lebanon, where Israel achieved lesser military success). This was so even though in both cases the campaigns were perceived in their respective countries as their main theatres of conflict and of vital national significance.

The Algerian and Israeli cases, as well as Ireland's, largely pull the carpet out from under another explanation for the puzzle: that developed powers' limited investment and much lower breaking points represent a lower interest and, hence, lesser motivation in the conflict than those of the indigenous force, a factor that ultimately decides the outcome. The argument suggests that the 'balance of resolve' has outweighed the 'balance of capabilities' in such unequal wars.[9] Most irregular conflicts have taken place in the undeveloped and developing parts of the world, whether in a colonial or post-colonial setting. Liberal economists have always stressed that maintaining colonies did not pay, and, in any case, the countries in question were and remain among the world's poorest. As a result, this argument goes, the developed powers' interest in them was low, and once the costs of staying mounted, the incentive to pull out became irresistible. It should be noted, however, that Ireland was not a faraway colony but had been an integral part of Britain for centuries. Algeria, too, was regarded by the French as part of metropolitan France. The possibility of retreat tore France apart and brought it to the brink of civil war, not least because it involved the uprooting and removal of European settlers who had lived in Algeria for generations. The same applies even more to Israel and the territories occupied in 1967, which constitute the core of the country historically and geographically.

The awakening of modern nationalism is widely perceived as a crucial factor in galvanizing indigenous resistance. However, although there is some validity to this argument, it fails to explain the much greater success of nonliberal powers in subduing others, including – as we shall see – societies in which nationalism was fully developed.

The liberal democracies' record of failure in counterinsurgency warfare is frequently attributed to the effects of television coverage. It should be noted, however, that Britain lost the struggle against Irish independence long before television, as was effectively the case with the loss of its empire in general. Similarly, the French lost their war in Vietnam before the advent of television that would later allegedly lose the war for the Americans in the same theatre. Even the French defeat in Algeria (1954–1962) effectively predated the age of television coverage. American television coverage of the war in Vietnam became negative only at a late stage, when it became clear that the United States had little hope of winning.[10] Television's transmission of the horrors and atrocities of war, like the effect of earlier mass media – the newspaper, radio, and newsreel – only reinforced a trend that was already strongly in evidence and was becoming ever-stronger as liberal sensibilities deepened.

The image of near invincibility that insurgency, guerrilla, 'asymmetrical' warfare has acquired stands in stark contrast to its often low military effectiveness.

Insurgents have rarely been able to defeat regular armies militarily, and they sustain far greater losses than they inflict, sometimes crippling losses. Nor is it accurate that modern affluent liberal democratic countries tend to lose wars against irregulars because of the democracies' inability to withstand protracted wars of attrition and their need to decide a war rapidly, as some scholars have claimed.[11] In both world wars, grinding attrition was actually the democracies' strategy of choice, whereas their rivals, Germany and Japan, sought rapid decision by lightning campaigns. In the Cold War, too, it was the liberal democracies that outlasted the Soviets in the protracted conflict of materiel and endurance. Indeed, eight years of war for the French in Algeria and eight for the Americans in Vietnam hardly constitute short struggles. Conversely, the brutality of authoritarian regimes often cuts insurgencies short, so that in most cases the authoritarian stamina need not be put to the test. Liberal democracies have tended to lose protracted irregular wars against far less developed societies because their self-imposed restrictions on violence against civilian populations have ultimately rendered their often-successful military operations futile. Only when a significant (liberal) portion of their publics realized that no decisive, war-ending victory was possible under these circumstances did they turn against the continuation of these wars. Unable to win, they made withdrawal the default option.

This is not to say that the balance of interests has nothing to do with the outcome. It is important to note that on the rare occasions when insurgents' demands go beyond withdrawal from their native territory, the balance changes. Irish resistance dislodged Britain from most of Ireland but has not been able to achieve a similar result in the predominantly Protestant parts. Palestinian resistance drove Israel out of most of the West Bank and the whole of Gaza, but when Palestinian suicide bombing intensified in Israel in pursuit of more far-reaching Palestinian goals that touched on the very existence of Israel, the latter successfully reoccupied the West Bank. Similarly, forced out of southern Lebanon by Hezbullah in 2000, Israel surprised Hezbullah in 2006 by the magnitude of its (mishandled) response to the organization's further encroachments. The Chechens drove Russia out during its liberal phase in the 1990s, but severe terrorism by Chechen extremists in pursuit of further demands led to a Russian reoccupation. Liberal democracies may find it very difficult to win even when pushed to the wall; but in such circumstances they, too, are capable of enduring a conflict and outlasting the insurgents by not losing. Although insurgents are often able to force far stronger liberal powers to withdraw from insurgents' home territories, they are unable to force capitulation.

Some critics do not agree that liberal democracies are far less ruthless than authoritarian and totalitarian powers. A study of inter-state wars in the period since 1815 claims that democracies have been at least as inclined as nondemocracies to target civilians, if not more so. The British starvation blockade of Germany in World War I and Allied city bombing campaigns against Germany and Japan in World War II are major examples. But indeed, as the same study emphasizes, such instances predominantly occurred in desperate major wars for survival – a category that excludes the overwhelming majority of counterinsurgency wars against weak non-state rivals. Furthermore, the study documents (though barely takes into account) a change in the overall trend after 1945, which greatly intensified during a period of growing liberalism after 1970, in which democracies have targeted civilians less than nondemocracies.[12]

Critics may argue that during the twentieth century democracies still wielded formidable instruments of coercion and pressure in counterinsurgency wars and were often quite brutal. True, atrocities, tacitly sanctioned by political and military authorities, or carried out unauthorized by the troops, have regularly been committed against both combatants and non-combatants. All the same, strict restrictions on the use of violence against civilians constitute the legal and normative standard for liberal democracies. And although many, probably most, violations of this standard remain unreported, numerous incidents have been exposed in open societies with free media, and are met with public condemnation and judicial procedures. Indeed, the effect of liberal public opinion and the media has been mightily reinforced over the past decades by the courts' increasing involvement in the policy process through judicial review, which in the view of some has made 'lawfare' almost as decisive as warfare. All these developments radically limit the liberal democracies' powers of suppression, judged by historical and comparative standards. The fact that not only a massacre such as that of My Lai during the Vietnam War but also the terror and intimidation practices at Abu Ghraib prison in Iraq evoke the most resounding outcry is an illustration of the continuous rise in the standards of conduct that liberal democracies apply to themselves.

Sceptics might also question the notion that ruthless brutality is the *sine qua non* of successful counterinsurgency suppression, on the grounds that it conflicts with the 'winning of hearts and minds' that has been posited as the key to success in the liberal democratic discourse. Indisputably, winning over at least the elites of conquered societies – through benefits, cooptation, and the amenities of soft power – has always played a central role in imperial 'pacification', as Tacitus (*Agricola*, 21) memorably described with respect to the taming of the barbarian Britons by Rome. And yet that velvet glove always covered the iron fist that had crushed local resistance mercilessly in the first place, and remained unmistakably in place as the *ultima ratio* of foreign control. The winning of hearts and minds continues to be an important component in successful counterinsurgency; but it has become the liberal democracies' indispensable guideline for the pacification of foreign societies only because they have practically lost the ability to crush such societies by force if the latter choose to resist. Furthermore, the unpleasant truth is that the 'winning of hearts and minds' is both very rarely achievable in a foreign country and amid an alien and often hostile population and prohibitively expensive, whereas ruthless suppression is both highly effective and cheap. The notion that effective counterinsurgency is costly obscures the issue that is at the heart of the matter: counterinsurgency warfare becomes expensive and ineffective only when alternatives to ruthlessness are sought. This problem is scarcely felt by nonliberal democratic powers.

The use of urban environments by insurgents is a striking demonstration of the point. Traditionally, insurgency flourished mostly in the remote parts of the countryside, in deserts, mountains, forests, and swamps. Urban environment constitutes a deadly trap against an enemy who has no scruples about indiscriminately felling buildings and setting cities on fire, as in the past, or razing them to the ground from a safe distance with artillery fire, as in modern times, with little regard for the inhabitants. This traditional rationale was demonstrated by President Hafez al-Assad of Syria, who in 1982 had whole neighbourhoods in the city of Hama destroyed with artillery fire as his army brutally suppressed a revolt by the Muslim Brotherhood, killing an estimated 10,000–25,000 of the city's population in three days. The Muslim Brotherhood was crushed so severely that it remained ineffective for nearly 20 years, throughout the presidency of the elder Assad. The Russian conduct in Grozny (1999–2000) has been similarly ruthless, leaving large parts of the city in ruins but practically eliminating Chechen resistance there. Indeed, since this essay was written, the Russian intervention in Syria further demonstrated the point, most notably in Aleppo, Syria's second largest city. A major stronghold of the rebels against President Bashir al-Assad's regime, it was pulverized and made to surrender by the ceaseless Russian aerial bombing that indiscriminately targeted the rebels and civilian population alike (2016).

By contrast, irregulars fighting against liberal democratic powers make the cities their bastions precisely because they are able to blend into and take shelter within the urban-civilian environment, while relying on their opponents to refrain from operating indiscriminately in these settings. Thus the devastation caused by Israel in the villages of southern Lebanon, from which Hezbullah fired rockets on Israeli towns and villages during the 2006 Lebanon war, created an outcry both in Israel and abroad. This was so even though Israel had issued warnings to the inhabitants to leave, and 'only' about 1,000 Lebanese were killed – many, perhaps the majority, members of Hezbullah. This reaction replayed itself during the Israeli repeated operations in Gaza, which Israel had been hesitant to carry out, despite the rockets routinely fired on Israeli civilians, precisely because of Gaza's dense urban population among which Hamas interwove itself in order to gain cover. In Iraq, too, the cities of the 'Sunni triangle' and Baghdad proved to be the greatest challenge for the Americans after 2003. Still, the Israeli case is special, because the irregulars who fight it do so not in faraway countries, thousands of kilometres away, but on its own boarders, only a few dozen kilometres from Israel's own population centres. As they do not see themselves bound by moral limitations, an unprecedented situation has been created, whereby Hamas, Hezbullah and the other organizations now use their towns and cities as safe havens and launching pads for missiles which they shoot at Israeli towns and cities.

This is the phenomenon of unintended consequences in its most paradoxical effect. The morally-driven effort to distance the civilian population from the harms of war has been exploited by guerrillas who do not abide by this standard to plant their warlike capabilities and activities within the civilian medium, thereby *increasing* civilian involvement. This is somewhat akin to the so-called poverty trap in the developed world, where entitlements intended to alleviate hardship might sometimes actually expand and perpetuate it by creating dependency or because of a cynical exploitation of the system.

The authoritarian-totalitarian record of success

It follows from our argument that nondemocratic powers ought to exhibit a substantially better record of success in curbing insurgency through the use of ruthless tactics from the late nineteenth century to the present. Needless to say, countless atrocities were committed in colonial settings by all imperial powers, irrespective of the regime. The Congo (Zaire), the private domain of King Leopold II of Belgium, was notorious for the cruel acts of coercion employed by the king's agents to force the natives to work for the rubber industry. The French brutally suppressed local resistance during their conquest of West Africa, while the British were almost as ruthless in 'pacifying' Kenya, destroying crops and huts and capturing livestock to force the locals to surrender.[13]

And yet, even by colonial standards, Imperial Germany's conduct in Africa was exceptional. In German Southwest Africa, today's Namibia, the Herero revolt (1904) was countered by a policy and strategy of extermination. Wells were sealed off, and much of the population was driven out to the desert to die, while the rest were worked to death in labour camps. Only 15,000 of 80,000 Herero survived. In German East Africa, today's Tanzania, the Maji Maji revolt (1905–1907) was similarly answered with extermination. A small force of 500 German troops destroyed settlements and crops so systematically that 250,000–300,000 natives died, mostly of starvation, more than ten times the number of those who had risen in arms. The once populous area became a wildlife reserve.[14] These were chilling demonstrations of the effectiveness of the old techniques of imperial suppression, which ultimately rested on the threat and practice of genocide.

The same spirit extended beyond Africa. To the bewilderment of public opinion in the democracies, Kaiser Wilhelm II, addressing the German troops departing to participate in the suppression of the Boxer Rebellion in China (1900), called upon them to be as merciless as Attila's Huns. And German attitudes and practices in Western Europe itself had also become increasingly distinctive. Alarmed by French mass popular resistance in the latter part of the War of 1870–1871, the Germans reacted with great brutality (although hardly more so than that recommended by the American Civil War veteran General Philip Sheridan, who told Bismarck's entourage: 'Nothing should be left to the people but eyes to weep with'). Harsh measures against civilian resistance were incorporated into German military manuals after 1871, and were given free rein in 1914 in widespread atrocities in Belgium, wherever the invading German troops met or imagined civilian resistance or sabotage. And if a harsh regime, as imposed by the Germans in occupied Belgium during World War I, was able to extract only partial cooperation from the Belgians, Nazi Germany's unbridled use of terror secured total compliance during World War II.[15]

Nazi Germany controlled the countries of occupied Europe and successfully harnessed them to its war economy.[16] The famous 'resistance' was minimal, increasing somewhat only with the growing signs of German defeat. Only in Yugoslavia and some occupied parts of the Soviet Union did partisans' resistance in difficult terrain prove sustainable. Yet there can be little doubt that had Germany won the war and been able to deploy more forces to these troublesome spots, its genocidal methods would have prevailed there, too. Imperial Japan was similarly able to subdue Taiwan (occupied in 1895), Korea (1905), and Manchuria (1931), as it very likely would have been able to accomplish throughout its 'East Asian Co-prosperity Sphere', including China, had the Japanese empire survived World War II. German and Japanese imperialism was broken in the two world wars by defeat at the hands of other great powers, rather than being dismantled under pressure from indigenous resistance in their colonies. There are no indications that resistance would have stood a chance of succeeding in *their* cases.

Whereas the German and Japanese imperial experiences may be regarded as too brief for the purpose of comparison, Soviet imperialism was much longer in duration. For nearly a century, the Soviet Union suppressed the peoples of the old Russian Empire – Russians and non-Russians alike – with far greater brutality than the tsars had ever employed. It did the same in the countries it had occupied in Eastern Europe during World War II, until the disintegration of the Soviet totalitarian system for reasons other than indigenous resistance or unrest. Only in desolate Afghanistan did the invading Soviet forces fail to curb local guerilla resistance during the empire's wane, under circumstances we shall shortly discuss. Much the same applies to China, whose suppression of Tibetan and Muslim nationalism is likely to persist so long as China retains its nondemocratic regime. Thus occupied countries, including industrially developed ones permeated with a strong sense of nationalism – such as proverbially nationalistic Hungary and Poland – were controlled by and incorporated into totalitarian empires with relative ease. They were highly susceptible to ruthless pressure, to the extent that occasional demonstration rather than the actual application of such pressure was usually sufficient to keep them under the yoke.

Nonliberal powers, in other words, were less often involved in imperial wars of suppression precisely because they were so effective at suppression. Resistance was unable to grow into insurgency because it was deterred before it flared up. Because modern nondemocratic empires were held firmly under control until they were either crushed in great power wars (as with Germany and Japan) or dismantled peacefully when the totalitarian system disintegrated (as with the Soviet Union), the sample of insurgencies is strongly skewed, as it overwhelmingly comprises struggles against liberal democratic powers. The vast majority of anti-colonial insurgencies took place against liberal democratic empires, but, as Sherlock Holmes noted, it is 'the dog that didn't bark' under the totalitarian iron hand that is the most conspicuous. In studies of states' war-proneness, it is often forgotten that the nondemocratic imperial peace rested on successful suppression and terror. The liberal democracies' greater involvement in 'extra-systemic', mainly colonial, wars must be viewed in this light.

Indeed, given liberal attitudes, not only 'Mao's way' – violent insurgency – but also 'Gandhi's way' – mass civil disobedience and demonstrations – were ultimately sufficient to force liberal powers out. Gandhi saw very clearly that Hitlerism was supremely violent and murderous, and that it possessed none of the scruples that inhibited liberal countries. Yet he still advocated nonviolence as a method against it. He advised the Jews to opt for mass disobedience against Nazi genocidal persecution, and later offered the same advice to the occupied nations of Europe, calling on Britain to embrace civil defiance against a German invasion, in preference over armed resistance.[17] This fatuous proposal for the application of his approach only highlights the actual, unique historical and geopolitical limits within which it was able to work, and succeed – that is, when directed against liberal democratic powers. Had India been ruled by Nazi Germany, Imperial Japan or the Soviet Union, Mr Gandhi would have disappeared after his first noted activity, with no one getting the opportunity to learn who he was. Acts of civilian disobedience would have been mercilessly crushed.

Sceptics might question the efficacy of authoritarian-totalitarian regimes in suppressing and crushing insurgencies. The Soviet Union, did, after all, fail to subdue Afghanistan (1979–1988), despite the brutal tactics that caused an estimated 1 million civilian dead and millions of wounded and refugees. Yet ruthlessness has always been a necessary but not sufficient condition for effective suppression. Historically, defeating irregular warfare waged by a backward and fanatical rival in a vast, desolate, and sparsely populated country has always been a difficult undertaking. All premodern empires struggled with this problem. It was largely in this context that the Soviet Union failed to win the war in that dream guerrilla country, Afghanistan, just as the British had failed to do in the nineteenth century, while easily controlling India. The insurgents' ability to take refuge across the border in neighbouring Pakistan significantly contributed to the Soviet failure, as it presently complicates matters for the Americans. Additionally, the Soviet failure also signalled deep problems developing within the totalitarian system that would shortly lead to its collapse, including a certain loss of nerve in employing the Stalinist-type brutality that was essential for the survival of such a regime.

Under Stalin, the Soviet Union evinced no scruples in resorting to the classical technique for eradicating popular resistance, deporting whole populations (including the Chechens) *en masse* from their homelands. Popular resistance in the Ukraine and the Baltic countries prior to and following World War II was similarly crushed by the harshest of measures, occasionally escalating into a strategy of extermination. Afghanistan was the exception – the *outlier* – in the Soviet imperial system. It is no coincidence that the secession of the former Soviet republics, as well as the insurgency in Chechnya, took place only after the breakdown of the Soviet system. Furthermore, whereas the liberalizing Russia of the 1990s was forced to give in to the Chechen insurgents, the Russia that has been turning in a more authoritarian direction has proved tenacious in curbing this resistance by ruthless means. The order on the scale is unmistakable: Soviet methods under Stalin were the most brutal and most effective in curbing resistance, while Putin's Russia constitutes an intermediate case, with a liberalizing Russia of the 1990s proving the least brutal and least effective. One reason for Russia's backtracking from liberalism was a growing sense in the country that the application of liberal norms would result not merely in the dissolution of the Soviet bloc and the Soviet Union but would also threaten the territorial integrity of the Russian Federation itself.

Like Nazi Germany (especially after the outbreak of World War II), wartime Imperial Japan, and Mao's China, Stalin's Soviet Union cared very little about opinion in the liberal West. In other times and cases, however, the power and wealth of the liberal sphere has exerted at least some measure of constraint on those less brutal and less inward-looking nonliberal regimes that felt sufficiently dependent on cooperation with the liberal countries to pay some heed to their sensibilities. Since the beginning of the nineteenth century, even nonliberal countries have been working in the context of an international system in which liberal countries and liberal public opinion carry weight and must be taken into account. Napoleonic France is an interesting early example, for it already operated within the constraints and norms of an enlightened Europe to which France had contributed so significantly and which the Empire proudly claimed to represent. Even that rare case of judicial killing under the First Consul, the abduction and execution of the Duke d'Enghien (1804) – which is trifling by the standards of twentieth century totalitarianism – was greeted with an outcry and condemnation both at home and abroad. Thus, despite the widespread atrocities committed by both sides during the savage war in the Iberian Peninsula (those by the French were graphically depicted by Goya), Napoleon did not resort to the semi-genocidal methods that had been used by the Romans in their long struggle to 'pacify' that same difficult arena, though the Spanish ulcer was causing his empire to hemorrhage. Like the various circles of hell, there are degrees of brutality, and there is a hell of a difference between them. Though Soviet methods in Afghanistan and Russian methods in Chechnya were brutal by Western liberal standards, they still fall far short of the genocidal methods used by a Hitler or a Stalin to curb resistance and crush insurgency.

Conclusion

Counterinsurgency warfare's record of failure turns out to be mainly the lot of modern liberal democracies – arising, in fact, from their noblest of traits. This is generally obscured by the West's self-criticism and the rhetoric of decolonization. For all of colonialism's abuses and brutality (as well as its often forgotten blessings), resistance to colonialism succeeded almost exclusively against liberal democratic powers. By contrast, authoritarian and totalitarian powers crushed resistance with an iron fist, as they continue to do.[18]

This is not to say that the democracies always lose counterinsurgency wars, while nondemocracies always win. Democracies can succeed in the rare cases when they are able to isolate the insurgents from the civilian population or when pressed to the wall. Weak authoritarian regimes have sometimes failed in counterinsurgency conflicts because of their weakness, but strong authoritarian regimes, and especially totalitarian ones, rarely fail. Other factors are surely involved in each particular case, but the record is heavily tilted against democracies in such struggles.

Even when the moral case is viewed as imperative in liberal eyes, as in humanitarian interventions and peacekeeping missions aimed at stopping mass killings, action is prone to fail, or simply be deterred, when at least one of the local sides opposes it and adopts persistent irregular warfare. After the fiascos of international intervention in Lebanon (1982–1983) and Somalia (1992–1994), nobody was willing to act in the cases of Rwanda, Darfur, and, indeed, the Syrian civil war.

Indeed, while many may be sympathetic to the success of the guerilla in colonial settings during the past century, circumstances are considerably different today. Rather than serving the cause of liberation and national self-determination, guerrilla tactics in today's world are all too often employed in the service of the most sinister and extremist causes. Furthermore, there is the ominous prospect of non-state organizations obtaining and using weapons of mass destruction (see chapter 4 in this volume). Thus the problem of successfully fighting irregulars has returned with a vengeance, and it remains as intractable as before. In all likelihood, if a nuclear device were ever set off in a major Western city, or a 'superbug' developed in a clandestine laboratory were released to cause mass death, democracies would react with far fewer constraints than they have in the past – adopting means far in excess of even the controversial post-9/11 measures.

Short of this doomsday scenario, real remedies to the democracies' problem in counterinsurgency operations are in short supply. In the first place, there is still an unfulfilled potential in the adaptation of high-tech warfare – which is far more attuned to targeting hardware than people – to the task of fighting irregulars. Both the United States and Israel have increasingly invested in surveillance, tracking, and robotic systems designed to detect and dig out low-signature irregulars operating in dense civilian environments. Second, the democracies try to cultivate indigenous allies, who not only enjoy greater local legitimacy than a foreign power and are more familiar with the local populations, but, one must admit, are also less constrained in their conduct. The European aerial support for the rebels against Gaddafi in Libya was tailored to avoid the kinds of military involvement, most notably in urban settings, at which liberal democracies are at their weakest. The mayhem that followed in post-Gaddafi Libya is a reminder of the many additional problems that beset humanitarian interventions. A similar strategy and many of the ensuing problems can be seen in the United States' cooperation with the local Northern Alliance during the American air-led occupation of Afghanistan; with the Sunni tribal leaders that led to the weakening of al-Qaeda in Iraq; and with local forces – most successfully the Kurds – against ISIS in both Iraq and Syria. The limited effectiveness of such strategies in many cases, and many disappointments, as well as the remaining moral dilemmas, are part of the deal. But as long as they can be made to work, stand-off war and war by proxy are vastly preferable to foreign intervention on the ground.

Finally, even unfriendly state regimes, which can be coerced, usually constitute a better option than no regime at all or a wholesale foreign intervention. Of course, when vital interests are involved and indigenous state authorities either do not exist or are unable or unwilling to enforce their authority, direct military action on the ground may still prove necessary, revealing the liberal democracies at both their best and weakest.

Notes

1 This essay was written with Dr. Gil Merom of the University of Sydney.
2 The thesis developed here was originally put forward by Gil Merom, *How Democracies Lose Small Wars: State, Society, and the Failure of France in Algeria, Israel in Lebanon, and the United States in Vietnam* (New York: Cambridge UP, 2003). Ivan Arreguin-Toft, *How the Weak Win Wars: A Theory of Asymmetrical Conflict* (Cambridge, England: Cambridge UP, 2005), in effect ends up corroborating Merom's thesis. What he terms a strategy of 'barbarism' turns out in his analysis to be the chief method of suppressing counterinsurgency. Indeed, his statistics indicate a sharp decline in successful counterinsurgency since the nineteenth century, correlating with the decline in the application of barbarism, particularly, as he admits, by democracies. See especially pp. xi–xii, 4, 33, 37, 204, 225.
3 David Edelstein, *Occupational Hazards: Success and Failure in Military Occupation* (Ithaca, NY: Cornell UP, 2008), is a recent example of the common blind spot for the role of democracy. As it apparently does not occur to him that the cause may lie precisely in this factor, all his case studies except for one involve a modern democratic occupier, which usually fails unless the occupier and occupied share enough common interests or common enemies. The only case study that involves a nondemocratic country, and could serve as a control, is that of the Soviet Union in North Korea in the immediate aftermath of World War II (and longer in Eastern Europe). The author concludes that in contrast to the US failure in South Korea during that same time, the Soviet occupier succeeded – indeed, easily and cheaply – because of its effective suppression capability.
4 Thucydides, iii. 37.
5 Harold Selesky, 'Colonial America,' in M. Howard, G. Andreopoulos, and M. Shulman (eds.), *The Laws of War: Constraints on Warfare in the Western World* (New Haven, CT: Yale UP, 1994), chap. 5.
6 Merom, *How Democracies Lose Small Wars*, p. 61.
7 Richard Shannon, *Gladstone and the Bulgarian Agitation 1876* (Hassocks, England: Harvester, 1975).
8 James Ray and Ayse Vural, 'Power Disparities and Paradoxical Conflict Outcomes,' *International Interactions*, 12 (1986), pp. 315–42.
9 Glenn Snyder, 'Crisis Bargaining,' in C. Hermann (ed.), *International Crises: Insights from Behavioral Research* (New York: Free Press, 1972), p. 232; Steven Rosen, 'War Power and the Willingness to Suffer,' in B. Russett (ed.), *Peace, War, and Numbers* (Beverly Hills, CA: Sage, 1972), pp. 167–83; Andrew Mack, 'Why Big Nations Lose Small Wars: The Politics of Asymmetrical Conflict,' *World Politics*, 27 (1975), pp. 175–200.
10 For revisionist reappraisals of the alleged critical role of television in Vietnam see: Daniel C. Hallin, *The 'Uncensored War': The Media and Vietnam* (Berkeley: U. of California P, 1986); William Hammond, *Reporting Vietnam: Media and Military at War* (Lawrence: U. of Kansas, 1998).
11 Dan Reiter and Allan Stam, *Democracies at War* (Princeton, NJ: Princeton UP, 2002), chap. 7.
12 Alexander Downes, *Targeting Civilians in War* (Ithaca, NY: Cornell UP, 2008), is thoroughly supported by statistics. Yet the author does not allow the clear change in the trend during the second half of the twentieth century, which he documents, to alter his general pronouncement regarding the somewhat greater propensity of democracies to target civilians. In a book concentrating on inter-state wars, he devotes one chapter to counterinsurgency warfare, and in it repeats his argument that democracies are no different with respect to the targeting of civilians. However, not only is his sole, early case study the Boer War in South Africa; he also fails to consider that just a few years later, Germany crushed insurgencies in both Namibia and Tanzania through genocidal

means. Indeed, although acknowledging that nondemocracies alone are prone to mass killings in the millions, Downes does not weigh numbers in his statistical analysis. He takes no account of the actual percentages killed and, moreover, of the guerrilla wars deterred or cut short by nondemocracies' threat of mass killings. All this creates a strong selection bias.

13 J. Moor and H. Wesseling (eds.), *Imperialism and War: Essays in Colonial Wars in Asia and Africa* (Leiden, the Netherlands: Brill, 1989), pp. 87–120 (esp. p. 106 for suppression techniques and atrocities), pp. 121–45 (esp. p. 141), pp. 146–67 (esp. p. 157).

14 Jon Bridgman, *The Revolt of the Hereros* (Berkeley: U. of California, 1981); Horst Drechsler, *'Let Us Die Fighting': The Struggle of the Herero and Nama against German Imperialism, 1884–1915* (London: Zed, 1980); John Iliffe, *Tanganyika under German Rule, 1905–1912* (Cambridge, England: Cambridge UP, 1969), pp. 9–29; G. Gwassa and J. Iliffe (eds.), *Record of the Maji Maji Rising* (Nairobi: East African Publishing House, 1967).

15 Geoffrey Best, *Humanity in Warfare* (London: Methuen, 1983), pp. 226–8, 235–7, and chaps. III–IV in general.

16 Peter Liberman, *Does Conquest Pay? The Exploitation of Occupied Industrial Societies* (Princeton, NJ: Princeton UP, 1996).

17 H. Jack (ed.), *The Gandhi Reader* (Bloomington: Indiana UP, 1956), pp. 317–22, 332–9, 344–7.

18 Michael Walzer, *Just and Unjust War* (New York: Basic Books, 1977), chap. 11, intelligently discussing the moral dilemmas involved, which in effect apply only to liberal societies.

8 The return of the authoritarian-capitalist great powers
Is the democratic victory preordained?

Does the liberal political and economic system have inherent advantages under modern conditions, which explain – indeed, 'guarantee' – its triumph, as Francis Fukuyama most famously, but also others, believe? The answer to this question is crucial both for understanding the causes of the democracies' victory over their rivals during the twentieth century and for assessing where today's world is heading. The case of the capitalist economy seems clearer: it has expanded almost irresistibly since early modernity, with the low-priced goods and the superior power that it produces working in tandem to erode and transform all other socio-economic regimes – a process most memorably described by Karl Marx in the *Communist Manifesto*. Indeed, contrary to Marx's expectations, capitalism had the same effect on communism, eventually 'burying' it without a shot being fired. But have political liberalism and democracy displayed a similar inherent advantage? Here the answer is far less clear. The triumph of the market, precipitating and reinforced by the industrial-technological revolution, has inextricably brought about the rise of the middle class, urbanization, spreading education, the emergence of 'mass society', and growing affluence. During the nineteenth century, in the 1950s and 1960s and again in the post-Cold War era, liberalism and democracy were widely assumed to emanate almost inevitably from these developments. But were there other viable alternatives which had the ability to keep in step with modernity? If so, why did such alternatives fail to prevail, and what does this past failure teach us about their future prospects?

Current attention focuses on two distinct challenges. One is posed by those traditional societies and cultures – most notably the Arab countries and Islam – that have so far failed to successfully embark on the road to the industrial-technological age, while also finding the hegemonic liberal order alien if not repugnant. However, although radical Islam is labelled as the new fascist challenge, the societies from which it arises are generally poor and stagnant. They represent no alternative model for the future and pose no military threat to the developed liberal democratic world, as did the fascist powers, which were among the world's strongest and most advanced societies. Only the potential use of nuclear and biological weapons – by states and, perhaps even more, by non-state actors – makes the threat of militant Islam significant (see chapter 4 in this volume). The second challenge emanates from the rise of new nondemocratic great powers in an international system dominated, since the collapse of the Soviet Union and communism, by the United States and a liberal democratic hegemony. These upcoming great and superpowers are revealed to be the West's old Cold War rivals, China and Russia, only now appearing in a reversed order of magnitude, and with authoritarian-capitalist, rather than communist, regimes.

Authoritarian/totalitarian-capitalist great powers played a leading role in the international system until 1945 but have been absent since then, and it is with their renewed challenge that this chapter is concerned. Re-examining the broad sweep of the past century, the chapter asks why the democracies came out with the upper hand, how 'inevitable' their victory was, and how indicative that past record is of future trends.

Why the democracies won the three great power conflicts of the twentieth century

The past record is largely dominated by the trial of great power military conflict. The liberal democratic camp emerged victorious from all of the three gigantic great power struggles of the twentieth century – the two world wars and the Cold War – overcoming its authoritarian, fascist, and communist rivals alike. But what was it exactly that accounted for this decisive outcome? It is tempting to look for its causes in the special traits of the opposing systems and trace it to intrinsic advantages of the liberal democracies. One may start with the democracies' international conduct. Did they, for example, more than compensate for their inferior repression capabilities abroad with a greater ability to elicit cooperation through the bonds – and discipline – of the global market system? This is probably true with respect to the Cold War, but does not seem to apply to the two world wars. Did the liberal democracies succeed because ultimately they always stuck together? Again this may have applied mostly to the Cold War, when the democratic-capitalist camp was in any case greatly superior, while also profiting from the growing antagonism within the communist bloc between the Soviet Union and China. During World War I, however, the ideological divide was much weaker. The Anglo-French alliance was far from preordained, being above all a function of the balance of power rather than the fruit of liberal cooperation. Only shortly earlier, power politics had brought these bitterly antagonistic countries to the brink of war and had made Anglo-German cooperation a strong possibility. Liberal Italy's departure from the Triple Alliance and joining of the Entente despite its rivalry with France was a function of that realignment, as Italy's peninsular location precluded conflict with the leading maritime power, Britain. During World War II, France was quickly defeated, whereas the right-wing totalitarian powers fought on the same side. Studies of democracies' alliance behaviour tally with these observations.

If it was not the structure of their international behaviour, was it then inherent domestic advantages – economic, social, and political – that gave the liberal democratic great powers victory in the three great struggles of the twentieth century? Did the liberal democracies, despite their strong initial reluctance to engage in war and lower levels of peacetime mobilization, ultimately prove more effective in mobilization? In reality, all of the belligerents proved highly effective in mobilizing their societies and economies for total war. Conservative and semi-autocratic Germany during World War I committed its resources as intensively as its liberal parliamentary rivals. After its victories during the initial stage of World War II, Nazi Germany's economic mobilization proved lax and poorly coordinated during the critical years 1940–1942. Well positioned at the time to fundamentally alter the global balance of power by destroying the Soviet Union and striding all of continental Europe, Germany failed because its armed forces were meagerly supplied with the military hardware necessary for a task that proved to be far more demanding than expected. The reasons for this fateful failure remain in controversy, but are at least in part attributed to structural problems of competing authorities inherent in Germany's totalitarian regime. More significantly perhaps, during the crucial period from June 1940 to December 1941 there was a widespread feeling in Germany that the war had been practically won. In any case, from 1942 on (when it became too late), Germany's highly intensified mobilization levels caught up with and surpassed those of the liberal democracies (though not their production volume, that is, that of the United States). Imperial Japan's levels of mobilization during World War II, and those of communist Soviet Russia, similarly grew higher than those of the liberal democracies by means of ruthless efforts.

Nor did the right-wing totalitarian regimes succumb in World War II because the democracies held the moral high ground, as Richard Overy and others have claimed. In actuality, these regimes proved more inspiring than the democracies, and their soldiers, if anything, fought more tenaciously. During the 1930s and early 1940s, fascism and Nazism were the exciting doctrines that generated massive popular enthusiasm, whereas the democracies stood on the defensive ideologically, appearing old and dispirited. While France collapsed like a pack of cards in 1940, Germany and Japan (and the Soviet Union) fought desperately to the last.

Only during the Cold War did the Soviet communist economy exhibit deepening structural weaknesses, made all the more evident when compared to the increasingly sophisticated and globalizing market economy. The Soviet system successfully generated the early and intermediate stages of industrialization (albeit at a frightful human cost), excelled in the regimentalized techniques of military mass production during World War II and kept abreast militarily during the Cold War. Yet, because of the system's rigidity and inherent lack of incentives, it proved ill-equipped for coping with the more advanced stages of development and with the more diversified economy of the information age. Ultimately, the Communist bloc practically dismantled itself, as both communist China and the Soviet Union, independent of each other, progressively found their system inefficient, almost irrespective of their militarized conflict with the capitalist democratic world.

By contrast, there is no reason to suppose that right-wing, capitalist, totalitarian regimes such as Nazi Germany and Imperial Japan would have proven similarly inferior economically, had they survived. The inefficiencies that typically arise in such regimes due to a lack of accountability and favouritism might very well have been offset by higher levels of social mobilization. Owing to their more efficient capitalist economies, the right-wing totalitarian powers, Germany and

Japan (again, particularly the former), would most probably have constituted a more viable challenge to the liberal democracies than the Soviet Union ever did; Nazi Germany was so judged by the Western powers before and during World War II. It should be noted that the liberal democracies did not possess an inherent advantage over Germany in terms of economic and technological development, as they did in relation to their other great power rivals.

So why did the liberal democracies win? In answering this question a distinction needs to be made between the democracies' left- and right-wing great power rivals. The communist world's weakness was indeed fundamental and structural. While the capitalist camp, which in the wake of 1945 expanded to include all the rest of the developed world, possessed much greater economic power than the communist bloc, the inherent inefficiency of the communist economies prevented that bloc from ever catching up despite its vast resources. Together the Soviet Union and China were potentially larger than the democratic-capitalist camp, and, had they succeeded economically, other countries would have followed their example. This is also illustrated in the staggering difference in development between North and South Korea. By contrast, the right-wing totalitarian powers were defeated in war because they came against a far superior but hardly preordained economic-military coalition that combined the liberal democracies and the communist Soviet Union (with the latter bearing the brunt of the war during the most critical years). In the final analysis, the left-wing totalitarian powers were defeated because of their economic inefficiency, whereas the right-wing powers were simply too small, consisting essentially of two powerful but medium-size countries: Germany and Japan.

Thus, contingent factors may have played the decisive role in the triumph of the capitalist liberal democracies over their right-wing authoritarian and totalitarian challengers. And the most obvious and decisive of these contingent factors was the United States. After all, it was little more than a chance of history that this scion of English liberalism would sprout on the other side of the Atlantic, institutionalize its liberal heritage with independence, and then expand across the most habitable territories of the Americas, thinly populated by tribal natives, while sucking in massive immigration from Europe. It was but a chance of history that a power of continental scale and by far the world's largest concentration of economic-military power was thus created. Obviously, the United States' liberal regime and other structural traits had a lot to do with that country's economic success (consider Latin America) and even with its size, because of its attractiveness to immigrants. Yet if the United States had not been located in a particularly fortunate and vast geographical-ecological niche, it would scarcely have achieved its great magnitude in population as well as territory, as Canada, Australia, and New Zealand demonstrate. And location, of course, while crucial, was not everything, but only one necessary condition among many for bringing about a giant and, indeed, *united* States as probably the paramount political fact of the twentieth century.

Thus, even if its liberal system was a crucial precondition for the United States' colossal growth, contingency was at least as responsible for the fact that it emerged at all in the newly discovered territories of the New World, and thereby would ultimately be there to save the Old World, as Winston Churchill put it. This massive concentration of power, that during the twentieth century always surpassed the next two great powers combined, decisively tilted the global balance of power in favour of its allies. The liberal democracies constituted a greater aggregate of resources than their rivals because of that crucial fact as much as because of their advanced economies (which, again, were no more advanced than Germany's). The victory of liberal democracy was anything but preordained in either 1914 or 1939, though it may have been more secure in 1945. Yet, if any factor gave the liberal democracies their edge, it was above all the actual existence of the United States rather than any inherent advantage of liberal democracy. This 'United States factor' is widely overlooked in studies of the victory of democracy during the twentieth century. Put differently, if it were not for the United States, the liberal democracies would most likely have lost the great struggles of the twentieth century. This is a sobering thought, making the world created by these struggles appear much more contingent – and tenuous – than unilinear theories of development and the Whig view of history as Progress would have us believe. If it were not for the size and might of the United States, the judgement of later generations on liberal democracy would probably have recalled the negative verdict on democracy issued by the fourth century BC Greeks, in the wake of Athens' defeat in the great power struggle of that time, the Peloponnesian War. As is well known but easily forgotten: history is written by the victors.

Democratization following development?

But the audit of war and conflict is, of course, not the only one that societies – democratic and nondemocratic – undergo. One should ask what the future trajectory of the right-wing authoritarian and totalitarian regimes would have been had they not been defeated by war. Would they not in time and further development have come to shed their former identity, as the communist regimes eventually did, and perhaps even embrace liberal democracy? Obviously, the old agrarian elites and the autocracies based on them could not have survived under modern conditions. But was capitalist-industrial Imperial Germany, for example, ultimately moving towards increasing parliamentary control and democratization? Or would it have developed an authoritarian-oligarchic regime, dominated by an alliance between the officialdom, the armed forces, and industry, as Imperial Japan did, despite a brief liberal interlude in the 1920s? And, of course, there was the option of right-wing totalitarian dictatorship, again most ominously represented by Nazi Germany. Would it have mellowed over time, and even liberalized and democratized? The latter options seem highly unlikely, especially if Nazi Germany would have won the war, which would have appeared as a resounding confirmation of its hideous ideology. As all these major historical experiments were cut short by war, their possible future development remains a matter of speculation. However, can the peacetime record of *other* right-wing authoritarian regimes, mostly concentrated in the post-1945 era, offer an answer?

Studies that cover this period show that democracies have been the most successful economically. At the same time, authoritarian-capitalist regimes are revealed to have been as, if not more, successful in the earlier stages of development, while tending to democratize after crossing a certain threshold in terms of economic (and hence also social) development. The East Asian 'tigers' are the most striking example, but this seems to have been a recurring pattern also in Southern Europe and Latin America. All the same, there might be something misleading about the attempt to deduce a general, unilinear pattern of historical development from these findings. The problem is that the dataset may be skewed, because since 1945 the enormous gravitational pull exerted by the United States and by the expansion of the liberal democratic orbit has bended patterns of development worldwide. As the authoritarian/totalitarian-capitalist great powers, Germany and Japan, were crushed in war and these countries were threatened by Soviet hegemony, they lent themselves to a sweeping process of restructuring and democratization. Consequently, smaller countries, if they did not want to embrace communism, remained with no rival model to emulate and with no powerful international players to turn to other than the liberal. Their democratization after reaching a certain level of economic development might be interpreted as the result of wholly internal processes. Equally likely, it occurred under the overwhelming influence of the Western liberal hegemony – political, economic, cultural, and ideological. Probably *both* domestic processes and external pressures had an effect. Presently, Singapore is the only example of a first-rate economy that still maintains a semi-authoritarian regime, though it, too, is likely to change under the influence of the liberal hegemony. But indeed, are large-scale, efficient, affluent and nondemocratic Singapore-like great powers possible, while proving more resistant to encroachments from outside?

Authoritarian-capitalist China and Russia

This question has become highly relevant again with the re-emergence in the system of new nonliberal giants, above all the formerly communist and fast industrializing authoritarian-capitalist China. Russia, too, is retreating from its post-communist liberalism and assuming an increasingly authoritarian character. Will these countries, under the dominant impact of the affluent parts of today's international system, ultimately converge into the liberal democratic range? Or are they big enough to chart a different course and challenge the hegemonic model, creating a new nondemocratic but economically developed and powerful Second World? Might they, for example, create a form of authoritarian-capitalist regime, where the officialdom, industrialists, and the military ally, and which would be nationalist in orientation, while participating with lesser or greater restrictions in the global economy?

Even in its current bastions in the West, the liberal political and economic order might be shaken – by a crushing economic crisis affecting the global trading system in the direction of greater national and regional protectionism, by a resurgence of ethnic strife in Europe, or by any other unforeseen development. (This was written and published before the outbreak of the economic crisis in 2007–2008, and before the recent resurgence of nationalism in Europe and the United States.) And if the Western liberal model becomes less appealing and more troubled in its core countries, how would this affect the global periphery – in Asia, Latin America, and Africa – whose conversion to that model is much more recent, incomplete, and insecure, and hinges largely on foreign influence?

Thus the emergence of right-wing authoritarian-capitalist giants in the international system represents in part a return of long absent players and in part a new and unfamiliar phenomenon. Unlike pre-1945 Germany and Japan, the novelty of present-day China is that it is both by far the largest player in the system in terms of size while enjoying spectacular economic growth since its embracement of capitalism. Thus China has switched from a left- to right-wing – and far more efficient – brand of authoritarianism. As it is rapidly narrowing the economic gap with the developed world, China is on the road to becoming a superpower while holding on to its authoritarian regime.

Although the rise of the authoritarian-capitalist great and superpowers does not necessarily herald a nondemocratic hegemony or war, it does seem to imply that the democratic hegemony in the system, achieved after the collapse of the Soviet Union, might be short-lived and that a universal 'democratic peace' may still be far off. The new capitalist-authoritarian powers can be as deeply integrated into the world economy as Imperial Germany and Imperial Japan were, rather than strive to achieve autarky as was the case with Nazi Germany and the communist bloc. China, though maybe not Russia with its humiliating legacy of a lost empire, may also be less revisionist than the territorially confined Germany and Japan were. Still, China, Russia, and their future followers may stand on the other side of a real political divide, with all the potential for suspicion, insecurity and conflict that this entails. Furthermore, China in particular is destined to hold considerably more power than any of the democracies' past rivals ever possessed, being both large *and* economically advanced.

Of course, it may still be the case that over time, through a combination of internal development, increasing affluence and outside influence, China will make a transition to democracy, while Russia will reverse its drift away from it. All that this article argues is that there is nothing in the historical record to suggest that such a transition to democracy by right-wing authoritarian/totalitarian great powers is 'inevitable', while there is a great deal to suggest that such powers have a far greater economic and military potential than their left-wing counterparts. If China (and Russia) do not turn democratic, it will be of crucial importance that India remain democratic, both because of its vital role in balancing China and because of the model that it presents for other developing countries. Finally and most significantly, for all the more or less justified criticism levelled at the United States, its existence as the greatest aggregate of power in the system – and its alliance with Europe, Japan, and, indeed, India – remains the single most important boost for the future of liberal democracy.

Despite problems and weaknesses, the US' position is not too bad. Its GDP and productivity growth rate are the highest in the developed world. Furthermore, as an immigration country with only about one-fourth the population density of the European Union and China, and one-tenth of the population density of Japan and India, the United States also continues to grow significantly in population, whereas all the others are experiencing aging and ultimately shrinking populations. China's economic growth rate is the highest in the world and, given its huge population and still low levels of development, its growth harbours the most radical potential for change in global power relations. However, even assuming that China's superior growth rate persists and it surpasses the United States in terms of total GDP by the 2020s, as is variably forecasted, China would still be left with just over one-third of the United States' wealth per capita, and, hence, with considerably less economic and military power. Closing that far more challenging gap in technology with the developed world, even if achieved, would take several more decades. As it was during the twentieth century, the United States factor remains the greatest guarantee that even if liberal democracy does not retain its present-day hegemony, it will not be thrown on the defensive and relegated to a vulnerable position on the periphery of the international system.

Criticisms and responses

In 'The Myth of the Autocratic Revival: Why Liberal Democracy Will Prevail' (*Foreign Affairs*, Jan.-Feb. 2009), Daniel Deudney and John Ikenberry take issue with my 'The Return of Authoritarian Great Powers' (*Foreign Affairs*, July-Aug. 2007). They restate modernization theory, according to which there is only one sustainable route to modernity. Those countries unfortunate enough to have developed along paths different from the one originally taken by Britain and followed by the United States eventually had to converge onto the liberal path, either because they proved inferior to the democracies in terms of power, or because of the intractable internal contradictions they were bound to experience, which ultimately caused their systems to implode or undergo transformation. Liberal democracy is presumed to possess intrinsic selective advantages, which confer an air of inevitability on the past as well as on the future, and give much cause for optimism. If 'world history is the world's court', as Hegel put it, then History's verdict appears clear-cut and the End of History is preordained. But is it really? Or might the owl of Minerva encounter optical illusions, and its current point of flight prove as transient as any other?

The past

Deudney and Ikenberry disagree with my suggestion that the defeat of Germany and Japan in the world wars was crucially affected by contingency, in effect claiming – with Hegel – that accidents are not that accidental but rather happen to those who are accident-prone. They cite Nazi Germany's critical production failure in 1940–1942, and other shortcomings in the war efforts of Germany and Japan, as indicative of deep-seated structural problems within the totalitarian systems that ultimately led to their defeats. Maybe, but one should guard against lopsided interpretations. Germany remedied its production failure from 1942 on. In World War I it had experienced no similar failure, nor did Japan suffer from extraordinary war effort failures in World War II. In both wars the capitalist nondemocratic great powers performed great feats and won shattering victories. On the other side, the democracies went from blunder to blunder in both world wars: their publics and governments were dangerously late to rise to the challenge and their armed forces were ill-prepared, particularly during the 1930s; their initial defeats were potentially catastrophic; and their conduct thereafter was anything but free from serious errors.

All sides, then, committed serious mistakes and occasionally blundered, sometimes for reasons relating to their respective systems. However, contrary to the comforting notion that the democratic system eventually proved superior, the main difference lay in the fact that the Axis' mistakes were simply *unforgiving*, given the relative size of Germany and Japan and the odds they faced. As these powers were smaller than their adversaries, their performance was less failure-tolerant. To break out of its limited territorial confines and fatally cripple the superior coalition assembled against it in either of the world wars, Germany critically required a consecutive string of major victories. Indeed, it came remarkably close to achieving that goal in both world wars, with particularly horrifying potential consequences had it won World War II, but also with major consequences for the future of liberalism and democracy worldwide had it won World War I. By contrast, the colossal power of the United States meant that the democracies were able to sustain catastrophic failures, including the fall of their Russian ally in World War I and the fall of France and shattering of the US Pacific fleet at Pearl Harbor in World War II – and still come back.

Indeed, if it were not for the existence of the United States, Britain and France would probably have lost to Germany in both world wars. We might have had a very different, and nondemocratic, twentieth century, a very different world today, and a very different story to tell by way of grand theories of development. The constructed grand narrative of the twentieth century would probably have underlined the liberal democracies' political divisiveness and decadence and the superiority of authoritarian/totalitarian cohesiveness, rather than the triumph of freedom. We are inclined to rationalize backwards, but the lessons of history are a tricky thing.

There is no other way to interpret Deudney's and Ikenberry's argument except that the victory of liberal democracy was virtually preordained. But in order to claim so, one would have to make the following assumptions:

a) That the existence of a huge liberal democratic United States as the paramount political fact of the twentieth century was preordained; that there was no way it could not have emerged, and in the form that it did, not even if the continent of America had not been there as an accidental fact of geography.

b) That there was no way Germany could have won either of the world wars in Europe; or alternatively, that having won, thereby vindicating its deep-seated anti-liberal traditions and establishing a *Herrenvolk* status in a pan-European empire, it would have *necessarily* liberalized, as would have Japan in the Far East.

It is hard to believe that anybody can seriously espouse either of these assumptions.

Twenty-first century challenges

We now turn from past to future, to the return of capitalist nondemocratic great powers. China is by far the most challenging case, which within a generation or two is projected to close the economic gap with the developed world.

Deudney and Ikenberry repeat the claim that nondemocratic regimes are necessarily ridden with corruption and cronyism that are bound to stall their development once they reach a certain level. But as Alan Greenspan, the former chairman of the US Federal Reserve Bank, writes in *The Age of Turbulence* (p. 275), Singapore, a nondemocratic first-class economy, is one of the *least* corrupt states in the world – as was Imperial Germany and its Prussian predecessor. It has become an axiom that corruption is inevitable in the absence of democratic transparency and accountability. Yet for Max Weber, Prussian-German bureaucracy became paradigmatic, and it was justly famous for its efficiency and clean hands. The secret of these model cases lies in the bureaucracy's high social status, strong ethics of duty and public service, and, in Singapore, high pay. Whether or not China's neo-Mandarin regime can establish similar standards remains to be seen.

The rule of law is surely essential for a well-functioning advanced capitalist economy, and the lack of such supposedly puts nondemocratic countries at a disadvantage. But Germany was semi-authoritarian until 1918, yet it had a very strong rule of law (and a first-class capitalist economy). The same applies to Japan until 1945, as well as to present-day Singapore.

If the economic argument in itself is not as conclusive as many believe it to be, might not the socio-political transformation generated by economic development be the cause for eventual democratization? Michael Mandelbaum, for example, argues that capitalism is synonymous with individual choice. People who become used to exercising consumer choice in every decision of their daily lives can be expected to demand the same right politically. Thus, nondemocratic-capitalist regimes are based on an internal contradiction that makes them inclined to implode.

There is much truth in this argument and it appears very convincing, until one is reminded that life is full of contradictions and tensions that do not necessarily implode. Capitalist democracy itself is a combination that has always been torn between the great economic inequality generated by capitalism (which also biases the democratic political process) and democracy's overwhelming egalitarian drive. This tension was so stark that socialists throughout the nineteenth and much of the twentieth centuries regarded it as an irreconcilable contradiction certain to doom capitalist democracy, leading them to ordain socialism – economic democratization – as the wave of the future. In the meantime, some of the tension has been alleviated through the institution of the welfare state in democratic-capitalist countries. And yet the tension continuously remains very close to the surface, occasionally bursting out. In real life, then, people regularly live with tensions and contradictions, and the question is which of these prove to be overwhelmingly irreconcilable.

Ronald Inglehart and Christian Welzel, 'How Development Leads to Democracy' (*Foreign Affairs*, March-April 2009), offer a value-centered version of modernization theory, based on their important comprehensive surveys of world values. They document clear differences between low- and high-income societies, with a shift occurring from the 'survival' values of traditional societies and their emphasis on religion, respect for and obedience to authority, and national pride – to individuality, self-fulfillment, and tolerant sexual mores in affluent, 'postmodern' societies. Based on twentieth century experience, Inglehart and Welzel reasonably argue that such transformation of values lays the groundwork for democratization. However, like other varieties of modernization theory, theirs too overlooks the fundamental question at hand: social values undoubtedly change with modernization; but are the modern values they record indeed an inevitable product of industrialization and greater affluence? Or, alternatively, has this particular set of values itself been decisively shaped by the overwhelming liberal hegemony that the United States and the West have exercised since the destruction of the capitalist nondemocratic great powers – with their strong group values – in the first half of the twentieth century?

Inglehart and Welzel stress the persistence of different cultural traditions and significant cultural variations even among societies that have undergone modernization. It has yet to be seen whether genuine alternatives to, or significant variations on, the prevailing hegemonic values and socio-political system compatible with modernity arise and prove viable, most notably in East Asia. In this region, the world's most populous and fastest developing, long and deep-seated cultural traditions emphasize the community, social order and social harmony.

Inglehart and Welzel are careful to note that the modernization leading to democratization process is not deterministic but probabilistic. Nonetheless, they leave the strong impression that all that is necessary for it to take its course is time. A propensity, however, even a strong one, is just that; whether it, rather than other propensities, will be realized depends on circumstances, countervailing forces, contingent events, and other imponderables.

A liberal international order or opposing systems?

We come to the question of how to deal with a superpower nondemocratic China in the international arena. There exists the prospect of mutually beneficial economic cooperation and peaceful coexistence, but the possibility of antagonism and occasional conflict cannot be ignored. Again Deudney and Ikenberry, as well as Inglehart and Welzel, exhibit undiluted liberal-internationalist optimism, parts of which are easier to agree with than others.

China's free access to the global economy fuels its massive growth, thereby strengthening it as a potential archrival, a problem encountered by nineteenth century free-trading Britain when it faced other industrializing great powers. According to Inglehart and Welzel, there is little to worry about, because rapid development will only quicken China's democratization. But Britain's great fortune was that its hegemonic status fell into the hands of *another* liberal democracy, the United States, rather than being taken over by nondemocratic Germany and

Japan, whose future trajectories remain uncertain at best. With respect to China, one option could have been to make access to the liberal global economy conditional on democratization. But it is doubtful that such a linkage would have been advantageous or even feasible. Not only has China's economic growth been highly beneficial to economic prosperity worldwide, making the developed world and the United States in particular as dependent on China as China is dependent on them; but economic development and interdependence in themselves – in addition to democracy – are also a major force for peace. The democracies' ability to affect internal democratization in countries much smaller and weaker than China is very limited, and pressure could backfire, souring relations with China and diverting its development onto dangerous paths.

On the other hand, Deudney's and Ikenberry's suggestion that China's admission into the institutions of the liberal international order established after 1945 and 1991 will oblige it to conform and transform equally fails to take stock of the realities. Large players are unlikely to accept the existing order as it is, and their entrance into the system is as likely to change it as to change them. For example, the Universal Declaration of Human Rights, drafted by Eleanor Roosevelt, Réné Cassin and other dignitaries, was adopted by the UN in 1948, in the aftermath of the Nazi horrors and at the high point of American liberal hegemony. Yet the Human Rights Commission, and the Council that replaced it, have long been dominated by China, Cuba, and Saudi Arabia and have a clear illiberal majority and record. There are endemic and possibly irreconcilable dilemmas for liberal democracies here, which must be fully recognized as such.

That said, the possibility of a protectionist turn in the system needs to be considered as well, and worked against as a top priority. The prospect of growing protectionism in the world economy at the turn of the twentieth century and the protectionist turn of the 1930s were largely responsible for a radicalization of the right-wing nondemocratic powers and the slide into both world wars.

Critics argue that, unlike liberalism, capitalist nondemocratic systems have no universal message to offer the world, nothing attractive to sell that people can aspire to, and hence no 'soft power' for winning over hearts and minds. But there is a flip side to the universalist coin. In the first place, many find liberal universalism dogmatic, intrusive, and even oppressive. Even in the West, most notably in the United States, there has been a conservative backlash against a perceived excessiveness of liberal ideology, and in other cultures the threat perception is much higher. Resistance to a 'unipolar' world concerns not only the power of the United States but also the hegemony of human rights liberalism. There is a deep and widespread resentment in non-Western societies to being lectured by the West, and to the need to justify themselves by the standards of the hegemonic liberal morality that preaches individualism to societies that value the community as the greater good. There is also pervasive dissatisfaction with hyper-promiscuity and unbridled hedonism, associated with Western liberalism. Compared to any other historical regime, the global liberal order is in many ways benign, welcoming, and based on mutual prosperity, so it is natural for people in the West to believe that everybody else would want to join it. And yet both Germany and Japan had to be literally pulverized before they were made to abandon some of their most cherished national traditions and join the liberal order.

Capitalist nondemocratic China and Russia offer not only a policy of non-interference but also a message of particularism, international ideological pluralism, state sovereignty, strong state involvement, group values, and indigenous cultural development. These are attractive not only to governments but also to peoples, as an alternative to American and Western dominance and as a counterforce to the sweeping blind forces of globalization. A message need not be formulated in universalistic terms to have a broader appeal, as the great attraction of fascism during the 1920s and 1930s demonstrated.

Deudney and Ikenberry label the suggestion that capitalist nondemocratic systems may pose a viable challenge to capitalist democracy the new orthodoxy. But surely this conventional rhetorical device could not be less appropriate here. I am not a prophet and I do not pretend to predict whether or not China will eventually democratize and Russia reverse its retreat from democracy. It is possible that liberal democracy's twentieth century triumphs have already spread its model so far and deep that the renewed challenge of the nondemocratic-capitalist great powers has come too late. But the opposite is also possible. A re-reading of twentieth century history which is more contingent, less unilinear and less triumphalist for democracy, should alert us to the need to keep an open mind and guard against the illusion that the evidence lends itself to any closed interpretation that that can be projected onto the future. The democratization of China and Russia and the ultimate triumph of democracy are far from preordained.

9 A compass to the Arab Upheaval

What can nineteenth century Europe teach?

The Arab Upheaval has been the cause of profound bewilderment in the developed world and among policy makers, not least in Washington. Great enthusiasm for the Arab Spring was quickly replaced by confusion and concern regarding Islamic democracy or an Islamist Winter, depending on one's perspective. This was as quickly supplanted by disconcert and despair in the face of military takeovers and ferocious civil wars. The European revolutions of 1848, the 'Spring of Nations', with their great hopes and dashed dreams, have often been cited as an analogue. But indeed, what can the European experience of modernization and regime change during the nineteenth and early twentieth centuries teach us about the contemporary Arab world? History does not quite repeat itself, as differences of conditions, place, and time are as significant as similarities. Still, history is the best we have got.

What makes nineteenth and early twentieth centuries Europe and the current Middle East akin is their relative position on the road to modernization. Here are some customary socio-economic indicators. According to the most authoritative estimates, by Angus Maddison, real terms GDP per capita in non-oil producing Arab countries is in the same range as mid- to late nineteenth century Europe (roughly one-tenth of today's affluent world). Urbanization rates in Egypt and Syria are, respectively, just below and above 50 per cent, a level crossed by Britain around 1850 and by Germany around 1900. Illiteracy in the major Arab countries still hovers between 20–30 per cent (greater among women than men), again in the same range as in mid-nineteenth century Europe (with the exception of the continent's highly literate northern countries). While these major indicators are of fundamental significance, differences should also be factored in. Whereas nineteenth century Europe and the West were the world's pioneers and world leaders in modernization, today's Arab countries are among the world's stragglers, with only Africa trailing behind. Because of this, the Arab world enjoys many of the fruits of modernization as imports from outside – in communications, household appliances, computers, medicine, and the like. This also means that the Arab world is susceptible to pressures from the hegemonic developed world – most notably economic, partly military, and, more ambivalently, intellectual – even if the efficacy of such pressures is inherently limited. Finally, there are all the differences of culture and historical traditions, for, as we know, the process of modernization, while most powerful and deeply transformative, is far from being unilinear.

In pursuing our comparison and analysis, the following key concepts will serve as our prisms: democracy, liberalism, development, religion, nationalism, and stability.

Democracy

The call for democracy has reigned supreme in the enthusiasm that surrounded the Arab Spring and the fall of the Old Regimes throughout much of the Middle East. It remains the strong expectation of Western opinion and the official demand by Western governments, most notably that of the United States. In today's West, democracy is perceived as the ultimate ideal and political norm, unconditioned by extraneous circumstances. But in reality, rather than democracy being an abstract, timeless idea waiting to be recognized and adopted by right-minded people, its successful implementation has always depended on and closely correlated with a number of developmental factors variably embedded in the process of modernization.

Thus it is anything but a coincidence that democracy on a large country-wide scale has never existed *anywhere* before modern times. And when it began to unfold in Europe and the West, even among countries that were turning democratic (and many countries were not), full democratization only arrived during the last decades of the nineteenth and first decades of the twentieth centuries. This was a protracted and gradual process. In France, for example, breakthroughs to democracy failed, first in 1793–1795, and again in 1848–1849. Democracy was only achieved with the Third Republic, from 1871–1875 on, while women's suffrage was delayed until 1945. In Britain, the majority of men were given the vote only with the Third Reform Act of 1884, and the expansion and equalization of suffrage among all men and women had to wait until 1918, 1928, and 1948. Developments in the Low Countries and Scandinavia were roughly similar. Even in the United States, with its unique foundations of liberty and relatively affluent estate of freeholders, practically universal white male suffrage was only enacted in the 1820s. Women were enfranchised in 1920, and black voting, despite the post-Civil War constitutional amendments, was secured only in the 1960s. Nonetheless, there is an implicit assumption that the rest of the world should profit from our experience and wisdom, skip all intermediate phases and leap right to the end.

As Aristotle observed, the middle class is the backbone of democracy, and its growth during the nineteenth century – hand in hand with industrialization, urbanization, improved communications, increasing affluence, and rising educational levels – constituted the material underpinning of democratization. True, very poor democratic countries also exist, most notably India after independence in 1947, and a much larger number of poor developing countries which democratized during the past 20 years, after the fall of communism and collapse of the Soviet Union. Still, other developing countries have proved highly resistant to democracy, whether for opportunistic or normative reasons. Moreover, there has always been not only an anti-liberal but also a liberal apprehension of democracy. Indeed, nineteenth century liberals were fearful of democracy, and imposed thresholds of property and education for voting, because they suspected that the masses, if enfranchised, would support neither democracy, nor liberalism, nor economic modernization. The same concerns are with us today, not least in the Arab Middle East.

The first concern is that a democratically elected government will not respect democracy. Going beyond despotism of the majority, this is popularly known as 'one man, one vote, one time'. The Jacobin government during the French Revolution, enacting universal male suffrage in 1793, but ruling dictatorially and ushering in a murderous reign of terror, is the modern precedent. Universal suffrage was abolished in the Constitution of 1795, after the fall of the Jacobins. 'One man, one vote, one time' became pretty much the norm in post-colonial countries in Asia and Africa during the 1950s and 1960s. It is the reality in Hamas-ruled Gaza, not to mention the Fatah-ruled Palestinian Authority in the West Bank, both since 2006. Fears that the Muslim Brotherhood in Egypt, which emerged as the largest party in the free parliamentary elections of 2011 and won the presidential elections of 2012, might take the same path did not materialize before the overthrow of President Mohamed Morsi and the legal steps against the party in 2013. Confidence in their popular support and a reluctance to spark an open clash with their opponents and the army were variably the reasons for the Brotherhood's relative restraint during their reign in power. Their opponents, however, proved less restrained in bringing down the Brotherhood's rule by force. More than abolition of electoral procedure by the Muslim Brotherhood's government, they feared its infringement on liberal rights and values.

Liberalism

Liberals everywhere in nineteenth century Europe were deeply concerned that democracy would jeopardize liberal rights, such as respect for human life, free speech, freedom of religion, toleration for a diversity of opinion and identity, and, above all, the right to property, which will be discussed separately. They feared that the masses would place little value on these hard-won socio-political norms, or else would be swayed by nonliberal creeds, whether traditionalist-conservative or revolutionary. A new age of barbarism was their nightmare, first predicted by Edmund Burke with respect to the French Revolution, which would indeed go down that road.

The Muslim Brotherhood's reign in Egypt was too brief to offer conclusive evidence, but the omens were not very good. The Brotherhood in power were relatively restrained, for the reasons mentioned. Nonetheless, they were ideologically and politically intolerant towards the large Christian minority, the Copts, and failed to respond to widespread incidents of violence against them. They held restrictive views regarding the role of women, and wished to impose their interpretation of Islamic modesty on the public sphere. They supported the practice euphemistically known as female circumcision, already widespread in Egyptian society. And they were sure to advance the incorporation of Islamic law, the *sharia*, into the country's legal code as much as they felt they could get away with. It is not surprising that many in Egypt, especially in the middle and upper-class neighbourhoods of the large cities, Cairo and Alexandria, felt threatened. And as in nineteenth century Paris and other European capitals, they were strategically positioned near the centres of power to take to the streets and make their influence felt, far more so than the masses of peasantry and supporters of the Brotherhood in the countryside. In Syria, the coalition of minorities – Alawites, Christians, Druze, and, partly, Kurds – that has supported and still supports the Assad regime is largely motivated by fear of the Sunni Muslim majority. This fear has only increased as the previously relatively moderate Sunni rebel movement is falling prey to militant Islamist and jihadist groups.

Western opinion and policy makers wish to see democracy installed and maintained, while also wishing that liberal values and norms be protected. They naturally tend to regard democracy and liberalism as inseparable, as the two have become in liberal democracies during the twentieth century. However, when the two sets of cherished values and norms conflict, which of them is to be given precedence? This question has long been absent from the script of Western and liberal democratic discourse. Moreover, liberal parliamentary regimes that were not democratic but later grew to become fully so were very much the norm in nineteenth century Europe. But their opposite, the recently posited concept of 'illiberal democracy', has rarely if ever materialized anywhere. The reason for this is that liberal values seem to be essential for a deep respect for a democratic system, as opposed to an opportunistic or instrumental attitude towards it. Illiberal democracies do not only infringe on liberal values and norms, but are also ever in danger of turning undemocratic too.

A major liberal right and concern is respect for private property, which interests us in its relation to economic development.

Development

The fear that the property-less masses would first move to confiscate the property of the rich was foremost in the minds of the bourgeoisie and liberals throughout nineteenth century Europe. Marx famously saw this as the reason why the 1848 Paris revolutionaries, fearful of the growing intrusion of the masses, surrendered the revolution to Louis Bonaparte, later Emperor Napoleon III. The popular, economically liberal, and variably authoritarian type of regime that he established, in the mode of his great uncle, is known as Bonapartism. The concept was applied to quite a number of rulers and regimes during the twentieth century, including, except for economic liberalism, that of Gamal Abdel Nasser in Egypt. It may partly apply to General Abdel Fattah el-Sisi in today's Egypt.

Marx had an economically developed communist society in mind, and his twentieth century successors experimented with socialist routes to industrialization and modernization. Thus private property was long a disputed value. However, the crushing failure of the nationalized economies pretty much left the capitalist route to modernity and affluence the only credible game around. In nineteenth century Europe, and in Britain even earlier, capitalist development was virtually imposed on the disenfranchised masses of peasantry and city poor, who saw little benefit for themselves from the process for generations, until much higher levels of affluence were achieved and the middle class swelled to become the majority. The most successful cases of economic development in the twentieth century, throughout East Asia, were similarly overseen by authoritarian regimes that imposed the process on all sectors of the population. The large majority of these countries eventually democratized after having achieved economic modernization. China is the latest giant of examples for this process at work, whereas democratic India is constantly obliged to placate its backward rural population, still the large majority of the population and a hindrance to more rapid modernization.

Economic development tends to be a prerequisite for a successful democracy, and the Arab overall record so far is one of abysmal failure. Unlike in other developing parts of the world, modernization in the Arab world has not taken off. The reasons for this are not easy to pinpoint. In the early twentieth century Max Weber singled out the cultures of both Confucianism and Islam as being detrimental to modernization and economic development. Since then, the spectacular development of East and Southeast Asia has often been credited to the virtues of Confucianism. Cultural traditions are more multi-faceted and adaptive than one assumed. The vogue of third-world socialism in its local form of Arab Socialism during the 1950s and 1960s was probably a major disruptive force for economic development in the Arab world. Unlike in India, in the main Arab countries it was coupled with and enforced by the ruthlessness of authoritarianism, which destroyed the urban commercial and entrepreneurial classes that had existed in Baghdad and Basra, Damascus and Halab (Aleppo), Cairo and Alexandria. Local Jews, Christian Arabs, Armenians, Greeks, and Italians, particularly active in this milieu, were pushed out. The result was two generations of economic stagnation. Before 1960, Egypt was better off than South Korea.

Who can be the agent of modernization in the Arab world and kick-start it on the long road from its currently nearly hopeless conditions? Might the change develop in the Arab monarchies, in countries such as Morocco and Jordan that have secured some modest achievements, enjoying traditional legitimacy and espousing development, while allowing limited electoral and parliamentary participation? Or can it be generated by an authoritarian or semi-authoritarian Arab Franco or Pinochet, or a Bonapartist of the type of General Sisi? None of these options looks very promising at the moment. Nor does a third one: a development-oriented, moderate and genuinely democratic Islamic movement. Unlike in nineteenth century Europe, socialism is dead in the Arab world. Nonetheless, there is a widespread popular resentment towards the rich, justified where corruption and crony capitalism are rampant, as they widely are, but otherwise detrimental to economic development. Much of this resentment is channelled into support for the Islamic parties, which preach virtue and social justice, and practise grass-roots social relief for the poor.

Religion

In some ways, political and social Islam resembles political and social Catholicism in nineteenth century Europe. Catholicism organized itself politically in reaction against the forces of secularism, modernity, liberalism, and democracy, preached non-worldly virtue and social justice, and practised social work for the poor. The most significant political party that exemplified the movement was the Catholic Center Party (*Zentrum*), which was consolidated to defend Catholic rights in Protestant-dominated unified Germany after 1871. Initially cast out by Bismarck, the party became increasingly integrated into the German political system, and became a major partner in the pro-democratic Weimar coalition after World War I. Destroyed by Hitler, it was replaced after World War II by the Christian Democrats who led Germany back from the abyss onto the road to democracy, liberalism, and economic development. Although it is customary to associate modernization with secularization, nineteenth century Europe saw growing religious piety, conformism and prudishness in many countries, for example in Victorian Britain, partly in response to the dislocation of traditional society and pressures of modernization. This was true of the middle class, as well as among the labour movement.

Can political Islam travel the same road and be transformed into the Arab and Muslim equivalent of the Zentrum and Christian Democrats? Even if unfamiliar with the historical precedent, this more or less was the hope of the US administration with respect to the Muslim Brotherhood government in Egypt. They wished to see a popular, broad-based, democratically elected movement that would increasingly learn to accept democratic procedures, make peace with liberalism and modernity, and embrace economic development. Potentially, this was the most attractive option. However, in Germany the process took generations to mature, as has also been the case with a closer and perhaps more relevant precedent: Turkey.

Modern Turkey is an especially instructive case. Shaped by Kemal Atatürk, with modernization as its supreme aim, it was a model of neither democracy nor liberalism. Kemalist Turkey made the army the guardian of a constitution that imposed secularism and banished Islam from the public sphere, probably against the people's majoritarian sentiments. Freedom of expression was curtailed on similar grounds, Islamic parties were suppressed, and the army repeatedly intervened in politics, removed democratically elected governments from power, and suspended democracy. Only in the 2000s was the mould of the Kemalist state broken, with the rise to power of the popularly based and greatly moderated Islamic Justice and Development Party (AKP). The party gave up its Islamic brand in favour of 'conservative democracy', is committed to economic development through the market system, and has come to terms with liberalism. The jury is still out on Turkey. But despite the authoritarian and fiery personality of the party leader and prime minister Recep Tayyip Erdoğan, large-scale liberal demonstrations against some of his initiatives and recent accusations of corruption, the new synthesis might be working, with the country making great strides in terms of socio-economic development. Still, 80 years of Kemalism may have laid the groundwork for this stage. They stirred Turkey away from a path similar to Pakistan's failed course, and avoided the major Islamist setback that followed the imperial modernizing hubris of the shah in Iran. Might Arab countries in time adopt a route similar to Turkey's, as Erdoğan recommended to the Islamists in Egypt?

Such a development might take time and require a preliminary semi-authoritarian phase, as it did in Turkey. The apocalyptic violent streak that Islamism has developed in recent decades is a major obstacle. So also is Islamic universalism and its challenge to the Arab states. Whereas militant violence was practically absent in nineteenth century political Catholicism (though not in other, revolutionary creeds), Catholic universalism was a much stronger reality. It nonetheless receded before the European nation-states, which were far more deeply rooted than their supposed counterparts in the Middle East.

Nationalism

The idea that national sentiments of affinity and solidarity, and their various political expressions, are exclusively modern and European is one of the great missteps of modern social theory. It is true, however, that mainly due to the dominance of imperial structures in the Middle East over millennia, nations and nation-states did not take root and evolve through most of this region. In the twentieth century, the new states that emerged were further undermined by the competing universalist ideas of pan-Arabism and the Islamic *ummah*. At the same time, they were undercut by the survival of tribalism, mainly associated with the existence of the pastoralist tribe in the semi-arid environments that do not exist in temperate Europe. Confessional communities, familiar from Europe, complete the cleavages and divisions of identity in the Arab Middle East.

Hence the Egyptian quip that except for Egypt itself all the other Arab countries are just tribes with a flag. Given their ethnic, confessional, and tribal disunity, many of these countries were held together only by the coercive and repressive force of brutal authoritarian regimes. Like the former Yugoslavia and Soviet Union, they have quickly fallen apart once these regimes lost power, sometimes degenerating into a murderous mayhem. Where a distinct national identity as one people exists, political differences, sometimes very acute and even violent, do not threaten the very existence of the state. Apart from Egypt, a people and nation with very old roots, national identity in the Arab world has more or less consolidated in Tunisia and partly also in Morocco and Yemen (where tribal politics are still very much alive). Among the non-Arab countries of the Middle East, Turkey, Iran, and Israel are all deeply rooted national states, despite the presence of large ethno-national minorities in each of them. By contrast, Syria and, only somewhat less ruinously, Iraq are the scenes of vicious violent conflict, torn apart as these countries are by their constitutive ethnic and confessional elements that often have very little in common with one another. In post-Gaddafi Libya, the various tribes and militias pay little heed to the central government and its flag.

Contrary to the academic cliché, nations are far from being easily 'manipulated' into existence from supposedly disparate communities in a process of nation building. Nor, contrary to American parlance, are the people in Iraq, the people of Iraq, and the Iraqi people interchangeable concepts. The same applies to Syria. Ethno-national differences arouse very deep human emotions and are politically highly potent and potentially explosive. The nineteenth and twentieth century disintegration of the Habsburg monarchy, the Ottoman Empire, the Russian Empire, the Soviet Union, and Yugoslavia are ominous reminders of the very bumpy road ahead in much of the Arab Middle East. Federalism, democracy, liberalism, and pluralism – all in short supply in the region – are the customary measures advocated in such cases. Sectarian violence, ethnic cleansing, and secessionist pressures are as much to be expected as uneasy coexistence.

Constructive optimism is what people tend to expect, but there is no guarantee of a good way forward or a happy ending. Up until the present, the first rule to learn about the Middle East is that good solutions have been few and far between and the realistic options generally ranged from bad to worse, with the primary question being which is the least bad. Might this change in the foreseeable future?

Stability

Nineteenth and early twentieth century Europe represents another ominous precedent. Socio-economic development, even when it finally took off, did not lead to democracy and liberalism everywhere. In many countries, including a highly developed and powerful Germany, the strongest counter-culture emerged in reaction against and resistance to Western liberalism. This mood found expression in a variety of authoritarian and semi-authoritarian movements and regimes, in some ways resurrected, with much of the same cultural baggage, in today's Russia and China. Fascism was the most extreme expression of that mood. In the Arab world, cultural hostility towards and resistance to Western liberal values and outlook are very deep and widespread and have always been so, since the beginning of Arab nationalism in the late nineteenth century. They have always coexisted with a grudging admiration for and humiliating sense of inferiority towards Western achievements, which have only aggravated the problem. The acutely felt gap between the manifest backwardness of the Arab world and the Arab-Muslim self-perception of their noble superiority is insufferable. These sentiments, simmering resentment and sense of cultural defensiveness are particularly powerful among the middle classes that have played a central role in the Arab Revolutions, quite contrary to their image in the Western media. Arab liberals are a tiny minority.

At variance with the progressivist view of modernization as ultimately spurring democracy and liberalism, only American power and crushing military victories in the two world wars shaped the world we know and take for granted. The twentieth century became the democratic century only *because* it was also the American century. Forced democratization actually did succeed in post-World War II Germany and Japan, but the prerequisites for this success need to be borne in mind: the two countries had first to be not merely defeated but pulverized in the war (and the communist and Soviet threat at their doorstep also helped), something which will hopefully never be repeated; both countries possessed the infrastructure of modern industrial societies, while also being cohesive national communities, attributes which are very much missing in the Arab Middle East. The future development of the countries in the region will depend to a large but limited degree on American and Western influence, both direct and indirect. But it may equally be influenced by the future trajectories of Russia and, most importantly, China, and on whether each of them will eventually democratize and liberalize politically. This cardinal question of the twenty-first century might also determine the viability of a potentially alternative model and source of support for Arab societies.

This article is not about concrete policy recommendations. But it does suggest that a sense of realism and proportion in assessing the odds and prospects of success is most needed in shaping American, and Western, policy towards the Arab Upheaval – present and future. While interventions, both military and nonmilitary, should not be ruled out, depending on the political, strategic, and humanitarian circumstances, their inherent limitations must be recognized. Given that the potential for hugely adverse developments in the region is at least as great as that for improvements, there is much to be said for preserving stability, as well as for patience and quiet support for a gradual fruition of local processes. Thus what appeared as high-handed American interference, more declarative than practical, against the ousting of the Muslim Brotherhood government in Egypt cannot really affect the process. It merely aggravates prevailing sentiments and deepens the cultural backlash against the United States and the West. It accentuates the notion among the Egyptian urban middle classes and other opponents of the Brotherhood that naïve American moralizing and disregard for the actual realities of Arab society requires them to surrender their freedom and way of life in the name of an abstract ideal of democracy. Saudi Arabia has a reactionary and in many ways objectionable social and political regime. But is a democratic and progressive alternative feasible at present or in the foreseeable future? Would a revolution that establishes a radical regime in a country that controls such a large share of the world's oil reserves be a better alternative? Has the regime of Khomeini and the ayatollahs in Iran been a better option than the shah, with all the flaws of his regime? Were the Bolsheviks and the untold horrors they inflicted a superior alternative to the reactionary and reprehensive Czarist regime, under which Russia was nonetheless beginning to experience accelerated industrialization and burgeoning parliamentarianism?

Needless to say, like nineteenth and early twentieth centuries Europe, the Arab world is not uniform or cut from the same cloth. As we have seen, different countries within the region have their particular characteristics and potentially different trajectories. Some give more cause for hope than others, and a discriminating understanding is called for. Still, like nineteenth and early twentieth centuries Europe, they share a great deal in terms of socio-economic development, history, cultural traditions (plus language), religion, and a sense of common identity. As the Arab Upheaval has demonstrated, they also deeply influence one another. The first obligation of doctors is to do no harm. Stability should not mean stagnation or be preserved at all costs. Democracy, even if imperfect, reasonably liberal values and norms, toleration, economic development, and internal peace should be cultivated as much as possible, coupled with awareness that outside influence has an inherently limited role to play. This is well recognized with respect to China and Russia. It no less applies to the Arab Middle East.

10 The Modernization Peace and twenty-first century conflict

The puzzle

Has the world become more peaceful? This notion first appeared in Europe during the nineteenth century, by far the most peaceful in European history until then. Yet it seemed to have been shattered by the two world wars, among the most destructive and lethal wars in history. They were followed by the Cold War, a clash of titans that arguably did not turn hot only because of nuclear deterrence. However, even before and increasingly after the end of the Cold War, new claims regarding the decline of war have been put forward. The find that democratic/liberal societies do not fight each other has progressively gained credence. An alternative theory has suggested that what we are witnessing is a capitalist rather than democratic peace. Others have detected an even broader phenomenon. In his *Retreat from Doomsday: The Obsolescence of Major War* (1989), John Mueller has claimed that the change was general, irrespective of political or economic regime, and that it preceded and was independent of the nuclear factor. My *War in Human Civilization* (2006), Steven Pinker's *The Better Angels of Our Nature: Why Violence Has Declined* (2011), Joshua Goldstein's *Winning the War on War* (2011), and Ian Morris's *War: What Is It Good For?* (2014), among others, have all argued that there has been a sharp decline in belligerency.

Thus, a number of questions arise. First, has war really been declining? Have the United States and its allies not been repeatedly involved in a series of messy wars over the past decades? Alternatively, is the relative peacefulness of today's world not attributed to a transient American hegemony since the collapse of the Soviet Union, to a fleeting post-Cold War moment? Are we not tempted once more by the old illusions that will yet again be dispelled by the rise of China to a superpower status, by a resurgence of Russia, or by vicious wars in south or central Asia, the Middle East, and Africa?

Second, if war has indeed declined, why has it? How valid is each of the explanations suggested for the decline? How do they relate to, supplant, or complement one another?

Third, when exactly did the trend begin: with the end of the Cold War, in 1945, or perhaps earlier?

Are nukes the answer? The long peace – not one but three

The good thing that was attributed to the nuclear balance of terror was its flip side: that the fear of mutual assured destruction (MAD) may have prevented the Cold War from turning into a third world war. As of now, the Long Peace among the great powers after 1945 has passed its 70th year, with the only partial exception being the limited war involving the United States and China in Korea (1950–1953).[1]

It has scarcely been noted, however, that the Long Peace among the great powers is not that special. It was preceded by 43 years of peace among the great powers between 1871 and 1914, with the Russo-Japanese War (1904–1905) as a partial exception. And this Second Longest Peace was preceded by the Third Longest Peace among the great powers, spanning 39 years between 1815 and 1854. Both the Second and Third Longest Peace occurred before the advent of nuclear weapons, and within a highly competitive multi-polar great power system.

However, are these long periods of peace unusual compared to earlier times? The most widely used database on wars and militarized disputes, the Correlates of War (COW), covers the period from 1816 onward. Thus, it has given researchers no indication of whether or not the Long Peace phenomenon is exceptional, or, indeed, no inkling that it exists in and is special to the period COW covers. One's horizons, and the questions that one is conscious of, have been constrained by the time span of the database. Fortunately, there exist a few longer range statistical studies of the trend.[2] A comparison of the European system before and after 1815 reveals that years of war among the great powers decreased by roughly two-thirds during the century following Waterloo, to only one-third of what they had been in the preceding centuries. Compared to their record during the eighteenth and seventeenth centuries, Austria and Prussia, for example – neither of them a democracy – fought about one-third to one-quarter as much during the century after 1815.[3]

There was no Long Peace between the great powers during the eighteenth century (or earlier). The following survey is restricted to the wars that pitted at least two of the five classical European great powers of the eighteenth and nineteenth centuries against each other: Britain, France, Prussia/Germany, Austria, and Russia. Even by this restrictive definition, there were only 2-year breaks between the various coalition wars of the French Revolution (and, later, Napoleon); 9 years between the end of the American Revolutionary War (which involved France in 1778–1783) and the French Revolutionary Wars; 15 years between the War of the Bavarian Succession (1777–1779) and the French Revolutionary Wars; 14 years between the Seven Years' War (1756–1763) and the War of the Bavarian Succession; 8 years between the War of the Austrian Succession (1740–1748) and the Seven Years' War; 17 years between the Great Northern War (which included Britain in the anti-Russian coalition in 1719–1721) and the War of the Austrian Succession; 6 years between the War of the Spanish Succession (1701–1713) and the Great Northern War. Great power wars during the sixteenth and seventeenth centuries were even more frequent.

Thus, the three Long Periods of Peace among the great powers are in fact highly unusual compared to earlier history. They have all occurred after 1815, and in a progressively ascending sequence, with the peace becoming ever longer each time. A clear trend had been building long before the nuclear age, beginning with the pacific nineteenth century. By the same token, however, one also needs to account for the massive divergence from the trend: the two world wars. A comprehensive account of the phenomenon in question must embrace both sides of the coin.

Has war become too lethal and destructive?

The first, most intuitive explanation for the decrease in major war is that the change must be due to modern wars becoming prohibitively costly in terms of life and wealth.[4] Statistical studies of the historical record from 1500 onward suggest that while wars became less frequent over time, they also became more severe, in the sense that more death and destruction were 'concentrated' in shorter spates of war.[5] Thus, perhaps there is no decrease at all in belligerency. Maybe a trade-off was created between the intensity and frequency of warfare: fewer and shorter but more cataclysmic wars supplanting a larger number of longer but less intense ones. However, as Jack Levy has found: 'the hypothesized inverse relation between the frequency and seriousness of war is not supported by the empirical evidence.'[6] In fact, for much of the period concerned, war has become less costly while its frequency declined.

This is most strikingly the case with respect to the century between 1815 and 1914. Not only did it experience both the First and the Second Long Peace between the great powers; but also, the wars that broke out during the period 1854–1871, interposed as they were between these two periods of peace, were far from the upper scale of the range in comparative historical terms. They included the limited Crimean War (1854–1856), the short Franco-Austrian War (1859), and the short and decisive Prussian-Austrian and Franco-Prussian wars (1866, 1870–1871). Nonetheless, following the last two wars, highly advantageous for the winner, Prussia, great power peace returned for another 43 years.

The world wars, particularly World War II, were certainly on the upper scale of the range in terms of the absolute number of war deaths (above 16 million in World War I; over 60 million in World War II), and they involved a staggering material outlay. However, absolute numbers are misleading for two reasons. In the first place, modern states are more populous and by far wealthier than their premodern predecessors, so the real question is what *percentage* of their people was lost in war and how much economic hardship war involved. Second, as their name suggests, the world wars, particularly World War II, encompassed a large number of states on several continents. Again, one needs to look at *relative* casualties, general human mortality in any number of separate wars that happen to rage around the world, rather than at the aggregate created by the fact that many states participated in the world wars. Absolute numbers distort our perspective.

In fact, the world wars are far from being exceptional in history. As Levy has concluded: 'According to the key indicators used here, the similarities between twentieth-century wars involving the Great Powers and sixteenth-to-eighteenth-century warfare are more profound than their differences.'[7]

Going beyond the modern European system, the vast expanses of world history tell as grim, and sometimes a grimmer, a story. Famously, statistics of any sort, including those of demographics and death, are extremely scarce if not entirely absent in the premodern record. Still, the cases outlined here come from some of the best documented civilizations and conflicts of premodern times. Taken together, they provide a rough but clear enough gauge, representing, like the world wars, some of the greatest and bloodiest wars in history.

The Peloponnesian War (431–403 BC) was ancient Greece's world war. According to Victor Hanson's calculations, the Athenian battle death toll reached about one-third of Athens' male population. If the famous plague is factored in – and it was directly caused by the war, as the rural population of Attica was evacuated into and was cramped within the city walls – perhaps one-third of the entire population perished.[8] This death toll is higher than that of any country in World War I and II *combined*. The economic cost of the Peloponnesian War was as staggering. None of this stopped the Greek world from continuing to fight within itself and against others later on. Similarly, we have some of the best of premodern manpower statistics in ancient Rome, which held regular censuses of its citizens. Here too, although the exact interpretation of the data is the subject of some controversy, the overall picture is clear enough. In the first three years of the Second Punic War (218–202 BC), the republic's most severe war, Rome lost some 50,000 citizens of the ages of 17–46, out of a total of about 200,000 in that age demographic.[9] This was roughly 25 per cent of the military age cohorts in only three years, in the same range as the Russian, and higher than the German, military death rates in World War II, and in a similar, if not greater, 'intensity' or 'concentration' over time. And the war went on for another 14 years. Indeed, the war's death toll and the devastation of the Italian countryside and of Rome's free peasantry during the war scarcely reduced Rome's propensity for war after the Second Punic War.

In the thirteenth century the Mongol conquests inflicted death and destruction on the societies of Eurasia that were among the worst ever suffered during historical times. The Mongol wars are also somewhat comparable to the world wars in their geographical extent. Estimates of the sharp decline registered by the populations of China and Russia vary widely. Still, even by the lowest estimates they were at least as great, and in China almost definitely much greater, than the Soviet Union's horrific rate in World War II of about 15 per cent.[10] If the Wikipedia List of Wars by Death Toll is any indication, and it is unlikely to be much off the mark, the Mongol conquests rank right after World War II as the most lethal in world history in absolute terms, and are much more lethal in relative terms than World Wars I and II combined.[11]

During the Thirty Years War (1618–1648) population loss in Germany, partly caused by war-related disease and famine, is estimated at between one-fifth and one-third[12] – again, either way, higher than the German death toll in World Wars

I and II combined. All these are some of history's greatest wars in terms of their scope and magnitude. But in countless smaller and less glorious wars between tribes, city-states, and states, relative mortality was often as high.

Nor have wars during the past two centuries been economically more costly than wars were earlier in history, again relative to overall wealth. War has always involved massive economic exertion and has been the single most expensive item of state spending.[13] Examples are numerous for both the destruction of and expenditure on war, and it will suffice to cite only a few instances, beginning with those we have already mentioned. The Mongol conquests wreaked such massive destruction on the highly developed civilizations of north and south China, Central Asia, Iraq, and Russia, that they took generations, sometimes centuries, to recover, and some of them never did. The same applies to the effect of the Thirty Years War on Germany. Both sixteenth and seventeenth century Spain and eighteenth century France were economically ruined by war and staggering war debts, which in the French case brought about the Revolution. Furthermore, as people in pre-industrial societies lived close to subsistence levels, expenditure on and the devastation of war quite literally took bread out of their mouths. Death by starvation and disease in premodern wars was widespread.

Did we simply learn to kick a senseless habit?

If the death toll, destruction of, and expenditure on war have not become greater after 1815, perhaps it is we who have undergone a change of heart. This begs the question of why the change now. According to Mueller, the 'attitude change' against war had no particular reason and was not different from a fashion or a fad that suddenly catches up.[14] However, whereas Mueller traces the 'attitude change' to the public backlash that followed World War I, the sharp decrease in war had in fact taken off much earlier, after 1815 and during the nineteenth century. From this period, Mueller still cites a succession of statements by philosophers, artists, and statesmen who extolled the noble virtues of war. It would thus seem that the attitude change trailed more than it prompted the change in reality.

Returning to the material calculus, perhaps war had *always* been unprofitable and has finally been recognized for the losing game that it is. Alternatively, maybe war has become unprofitable under modern conditions. If there has been no substantial change on the cost side of modern war, perhaps the change has taken place on the gain side. As Kaysen has put it in his critique of Mueller: perhaps war became unprofitable *before* it became unthinkable.[15]

According to a long-held view, war is fundamentally counterproductive and irrational, resulting from stupidity or from prisoner's dilemmas of various sorts which means that everybody loses from war and would be better off without it. However, while war does mean an overall net loss in life and wealth so long as it rages, the victors often secured control over a greater share of the rewards in dispute, from which they might continue to reap the benefits in the long run. While many wars ended in mutual loss, others brought huge gains to the victors.

Might the rationale of loss and gain have been somewhat different, in that the rulers and elite reaped the benefits of war whereas the common people were coerced into it against their interests and paid the price for it in life, property, and misery? It makes sense to think that as the people have become free and sovereign during the past two centuries they have increasingly refused to participate in the old game and shoulder the burden of war. This was in fact anticipated by Enlightenment critics of the Old Regime, such as Jean-Jacques Rousseau, the Marquis de Condorcet, Thomas Paine, and Immanuel Kant.[16]

However, as Alexander Hamilton had argued in the *Federalist Paper* (no. 6) in rejection of the pacific view of republicanism, a glance through history would have shown that some participatory republics were among the most bellicose, aggressive, and militarily successful states ever. It was not only the Chinggis Khans of history that profited hugely from war and eagerly pursued it. The Enlightenment thinkers believed that selfish autocrats and military aristocracies were responsible for war. And yet it was the populace in Athens, not the aristocrats, who pushed for aggressive imperial expansion and war. This was so despite the fact that the demos fought in the army, manned the rowing benches of the Athenian navy, and had to endure war's destruction and misery. Similarly, the secret of Republican Rome's extraordinary expansion and power was its participatory-inclusive regime which successfully co-opted the populace for the purpose of war and made mass citizen armies possible.

Why, then, did the citizens of Athens and Rome repeatedly vote for war and endure the death and devastating of protracted wars for years and years? It was because the gain side of war for them was even greater than its cost side. This gain included booty; lavish tributes (which in the Athenian case financed about half of the Athenian budget, paying for the extensive public construction and huge navy, in both of which the people were employed); trade monopoly; and massive land allocation in colonies established on territory confiscated from defeated enemies. Indeed, the more the people held political power and shared in the spoils of war, the more enthusiastically they supported war and imperialism and the more tenaciously they fought.

But wait. Both Paine and Kant suggested that war could be eliminated within a system of democratic/republican states. Perhaps democracies or republics fought tenaciously against nondemocratic states, but not against other democracies or republics.

The scope and limits of the democratic peace theory

The democratic peace proposition, corroborated by massive qualitative and quantitative research, suggests that democracies very rarely or hardly ever fight each other.[17] However, some fundamental questions with respect to the inter-democratic peace remain. One puzzling question concerns the non-application of this peace to premodern times. Thus, democratic Athens of the fifth century BC tyrannically ruled over an empire that consisted of democratic city-states, and mercilessly suppressed both aristocratic and popular revolts. Similarly, Athens launched an ultimately disastrous military expedition against democratic Syracuse (415–413 BC).[18] Half a century later, as democratic Thebes became supreme – defeating Sparta, setting its satellites and slave helots free, and establishing democracies everywhere – democratic Athens balanced against Thebes. It allied itself against Greek freedom with oligarchic and oppressive Sparta, its oligarchic allies, Greek tyrants, and autocratic Persia. Furthermore, as Athens' conduct towards the democratic members of its recreated empire again became oppressive, it prompted a rebellion known as the Social War (357–355 BC). Lest Thebes' conduct towards other democracies be considered saintly, remember that Thebes conquered and razed to the ground its old rival, democratic Plataea (373 BC).

Surprisingly, the record of Republican Rome's wars in Italy has not been examined at all in this context and appears to be no less questionable with respect to the democratic peace theory. For example, Capua and Tarentum, the two leading city-states of southern Italy that defected from Rome during the Second Punic War and were harshly crushed by it, were both democratic republics at the time.[19]

But if the inter-democratic peace did not apply before modern times, why did it not apply? The difference in this respect between premodern and modern democracies may provide a vital clue to the key element that makes the inter-democratic peace work.

A liberal or capitalist, rather than democratic, peace?

Some scholars have held that the peace in question is liberal rather than democratic.[20] They have claimed that the classical democracies can hardly be considered liberal, because they practised slavery and in general did not uphold the liberal rights and other republican preconditions required by Kant, such as a separation of powers.

The existence of slavery is indeed a major difference between ancient and modern democracies. Yet in what way exactly it is supposed to have affected foreign affairs is far from clear. The most common answer is that the institution of slavery generally meant less respect for human rights, equality, and dignity, which spilled over to relations between states. This again makes the modern peace a fundamentally ideological phenomenon. Materially, while slaves were certainly one of the main booties of war, it is difficult to make the case that slavery was that crucial among the many spoils of war, ancient or modern. As for liberal rights in general, freedom of opinion and speech, legal safeguards of life and property, due judicial procedure, and the rule of law were all central norms in ancient democracies and republics, even if in somewhat different forms than in their modern counterparts. Finally, whereas a separation of powers in the sense formulated by Montesquieu did not exist in antiquity, a separation of authority and checks and balances very much existed. Whereas classical Athens has become proverbial for the perils of direct democracy and tyranny of the majority, in the mixed-regime Roman republic institutional constraints were very strong. A popular assembly of the people in arms (*comitia centuriata*), the men from whom the legions would be raised, was called to vote on war, but only after the senate debated and decided on the question and a motion for war was introduced in the assembly by a consul who was elected annually in highly competitive elections.

It is evident, however, that liberalism as a distinct concept and major force is a creation of modernity. But if so, why? Again, this fact may harbour a vital clue. So far we have been concerned with the political aspect of liberalism, but political liberalism famously came hand in hand with economic liberalism both as a doctrine and a reality. Political liberalism emerged in the era of commercial capitalism in the seventeenth and eighteenth centuries, first in the Dutch Republic and then in Britain. Liberalism became yet more entrenched in Britain during the nineteenth century, when the country inaugurated industrial capitalism.[21]

Indeed, the prophets of the democratic-liberal peace during the Enlightenment – Montesquieu, Paine, and Kant – emphasized the economic dimension of that vision side by side with the political aspect.[22] The notion that mutual trade fosters peace ran through the nineteenth and twentieth century and has been more recently investigated extensively.[23] Some scholars have gone further, suggesting that what is known as the democratic peace is in reality a 'capitalist peace'. Erik Gartzke, analyzing the global data for the period 1950–1992, has found that mutual trade and common interest in the global markets is the most significant factor that reduces belligerency between states, whatever their regime. He has claimed that democracy in itself is a far less significant factor, but has conceded that the near absence of war between democracies stands out and remains to be explained.[24] Critics have found significant errors in Gartzke's statistical work, which after correction further reinforces the effect of democracy and the inter-democratic peace.[25] Moreover, the limited time frame of his study, confined as it is to the period after World War II, raises broader historical questions.

First, there is the problem of World War I, which pitted developed capitalist great powers closely integrated by mutual trade against each other. This was most conspicuously the case with Britain and Germany, the world's leading traders, which were each other's main trade partner (except for their imports from the United States).[26] The problem has been addressed by Patrick McDonald, whose work covers a much longer historical time frame than Gartzke's, spanning as it does both the nineteenth and twentieth centuries. McDonald has refined the capitalist peace concept, specifying free trade, competitive domestic markets, and a small government share in the economy as its cornerstones. He has pointed out that although the volume of international trade in relation to GDP was indeed very high in the first global age, before 1914, this was primarily caused by falling transportation costs and despite growing protectionism and high, rising tariffs from the later part of the nineteenth century. By this analysis, 1914 should not be regarded as a compelling counter-example to the trade-peace connection, because free trade did not exist.[27] We shall return to this point later on.

A second problem with the capitalist peace theory is that it overlooks the peace that existed among democracies between the two world wars, when international trade reached its modern nadir. In the wake of the 1929 financial crash, all the great powers, including the democratic ones, introduced strict protectionist policies. During that period, the United States' economy, for example, was seriously hurt by the tariff walls built around the British Empire, as Britain abandoned free trade and adopted protectionism from 1932 onward. The United States itself, which maintained high tariffs during the 1920s, raised them further in 1930. The result was a disastrous freefall in international trade, to only one-third of its pre-crisis levels. Nonetheless, although the Roosevelt administration put diplomatic and economic pressure on other countries to lower tariffs, the possibility of tearing down trade walls by the threat, let alone the exercise of force, was unthinkable. Indeed, this was despite the fact that American capitalism was constrained not only abroad but also at home, with the New Deal. Thus, the democratic peace worked during the 1930s, when the capitalist peace should have utterly failed.

Third, both Gartzke and McDonald fail to specifically address the communist great powers of the twentieth century, the Soviet Union and China, which totally eschewed capitalism, private property, and the markets, and were closed onto themselves in terms of trade. Nonetheless, these countries participated in the general decrease in belligerency, compared to their premodern record (particularly Russia's in the highly competitive European and Near Eastern systems). Indeed, this decrease is neither covered by the democratic peace, nor does it fall under the capitalist peace. Finally, what is to be made of social democratic countries? These primarily include the countries of Western Europe and, most conspicuously, the exceptionally peaceful Scandinavian countries, which while capitalist in many ways are characterized by deep state involvement in society and the economy.

Michael Mousseau has advanced another version of the capitalist peace, positing impersonal contract abiding cultural norms through the market rather than individual free enterprise as the hallmark of that peace. His concept could potentially accommodate social democracy, as well as New Deal America and other protectionist capitalist democracies of the 1930s.[28] However, it hangs everything on a very narrow feature, raising suspicion that it reflects a random statistical correlation or, more likely, an epiphenomenon of something more fundamental.[29] Were Paine and Kant perhaps more accurate in their multi-factor framework that encompassed mutual liberal republicanism, reciprocal trade, and shared international institutions?[30] This framework has proved to be so extraordinarily prophetic that we tend to overlook its shortcomings. Let us take stock of the gaps that we have seen: premodern democracies and republics actually did fight each other; nondemocratic great powers shared in the general reduction in belligerency during modern times, from 1815 on, including communist powers that largely opted out of the global trade system; and until the nineteenth century states tried to monopolize trade by force and bar all others out rather than share with them.

What then has changed from the early nineteenth century onward that can account for all of these factors? What was it that made states fight less, altered the preferences of democracies to a degree that practically eliminated war among them, and sharply increased international trade while reducing protectionism?

What is the missing element that can reconcile these observed phenomena, while encompassing and unifying the general decrease in belligerency, the modern inter-democratic-liberal peace, and the capitalist peace?

The Modernization Peace

The correlation between the decline of war and the process of industrialization and modernization – both unfolding since the early nineteenth century and practically overlooked in IR theory – is unlikely to be accidental.[31] But what is the causal relationship between them? Quite a number of individual elements within the modern transformation appear to variably correlate with the decrease in belligerency, as is the case with the nuclear peace, the democratic peace, the capitalist peace, and more. However, as suggested here, we are dealing with a composite effect. Some elements of the modernization process have affected the general decrease in belligerency from the very beginning of the industrial age, while others have emerged and added their weight later on during the past two centuries, as the process of modernization has unfolded. The combined deepening and accumulation of the factors involved tallies with the progressive decrease in the occurrence of armed confrontation (again, except for the spike during the world wars). Thus, the Modernization Peace should be comprehended in its entirety, as the previously mentioned factors would appear to be mutually interacting and mutually reinforcing aspects of a broader, more comprehensive phenomenon or development.

We begin with a snapshot of the distribution of war in today's world. Table 10.1 is a list of countries that saw inter- or intra-state armed conflict (at least 1,000 dead) taking place during the post-Cold War era. The conflicts are tabulated according to both region and wealth, measured by nominal GDP per capita (global ranking/2013 dollar value; oil producing countries marked with an asterisk).[32]

Some major features of these armed conflicts stand out. The world's 'zone of war' is located in the poorer parts of the globe, mostly in countries with GDP per capita in the lowest 50 per cent of countries (less than 4000 in 2013 USD). Wealthier countries that experience war on their territory are mostly major oil producers (marked by an asterisk). The reason for this is not necessarily that oil encourages war, but that the income from it gives a false picture of these countries' level of modernization.[33] Poor countries in general are particularly exposed to intra-state conflict, by far the most common type of armed conflict in today's world. This reality belies the possible claim that wars take place in the poor parts of the world because the rich countries are powerful enough to take them there and avoid war on their own territory. Although this claim is valid in a limited sense, it is grossly misleading overall, as most armed conflicts take place either within or between the less developed countries themselves, whereas no such conflict takes place either within or between the top 30 per cent of wealthiest countries (roughly above 13,500 GDP per capita in 2013 USD). This indicates a strong correlation between modernization and peace, and suggests that the armed conflicts that developed countries have with less developed rivals are a function of factors relating to the latter's poor record of development (Figure 10.1).

Table 10.1 Scenes of Intra- or Inter-State War in the Post-Cold War Era

By Region	By Wealth
Sub-Saharan Africa:	**From the 10th Decile (eight countries)**
Burundi (193/229)	Burundi (193/229)
Democratic Republic of Congo (192/286)	Democratic Republic of Congo (192/286)
Ethiopia (191/354)	
Liberia (189/356)	Ethiopia (191/354)
Eritrea (185/507)	Liberia (189/356)
Mozambique (181/579)	Eritrea (185/507)
Somalia (177/600)	Mozambique (181/579)
Rwanda (176/620)	Somalia (177/600)
Sierra Leone (169/725)	Rwanda (176/620)
Chad (167/818)	
South Sudan (161/1045)	**From the 9th Decile (seven countries)**
Sudan (151/1438)	Afghanistan (173/683)
Nigeria (129/2966*)	Sierra Leone (169/725)
Angola (98/5668*)	Chad (167/818)
	Tajikistan (162/1036)
	South Sudan (161/1045)
Middle East (Including North Africa):	Burma-Myanmar (158/1183)
Yemen (152/1422)	Pakistan (157/1238)
Syria (147/1606)	
Palestine (131/2908) (conflicts with Israel 29/37704)	
	From the 8th Decile (four countries)
Algeria (100/5325*)	Yemen (152/1422)
Iraq (97/5790*) (American & allies involvement)	
	Sudan (151/1438)
Lebanon (73/9793) (Israeli and American involvements)	
	India (149/1548)
Turkey (66/10972)	Syria (147/1606)
Libya (63/12029*) (NATO involvement)	
Kuwait (14/52198*) (Invaded by Iraq 97/5790)	
	From the 7th Decile (seven countries)
South and Southeast Asia:	Philippines (134/2765)
Afghanistan (173/683) (American & allies involvement)	Palestine (131/2908)
	Nigeria (129/2966*)
Burma-Myanmar (158/1183)	Sri Lanka (125/3159)
Pakistan (157/1238)	Guatemala (122/3478)
India (149/1548)	Armenia (121/3504)
Sri Lanka (125/3159)	Georgia (118/3715) (conflict with Russia 55/14680)
Philippines (134/2765)	

(*Continued*)

Table 10.1 (Continued)

By Region	By Wealth
East Timor (107/4362) (conflict with Indonesia 123/3475*)	Russia 55/14680*)

Latin America:
Guatemala (122/3478)
El Salvador (115/3826)

Peru (90/6593)
Colombia (79/7826)

Mexico (69/10293)

Former Soviet Union and Soviet Bloc
Tajikistan (162/1036)
Armenia (121/3504) – Azerbaijan (80/7814*)
Georgia (118/3715) (conflict with Russia 55/14680)
Ukraine(113/4024) (involving Russia 55/14680)
former Yugoslavia (Serbia 93/6313; Kosovo 128/2972)

From the 6th Decile (four countries)
El Salvador (115/3826)
Ukraine (113/4024) (involving Russia 55/14680)
55/14680*)
East Timor (107/4362) (conflict with Indonesia 123/3475)
Indonesia 123/3475*)
Algeria (100/5325*)

From the 5th Decile (four countries)
Angola (98/5668*)

Iraq (97/5790*)

former Yugoslavia (Serbia 93/6313; Kosovo 128/2972Kosovo 128/2972)
Kosovo 128/2972)

Peru (90/6593)
Azerbaijan (80/7814*)

From the 4th Decile (five countries)
Colombia (79/7826)
Lebanon (73/9793)
Mexico (69/10293)
Turkey (66/10972)
Libya (63/12029*)

From the 3rd Decile
Israel (29/37704) (conflicts with

Palestine 131/2908, and Lebanon 73/9793)

From the 1st Decile
Kuwait (14/52198*) (invaded by Iraq 97/5790)
97/5790*)
USA (12/52,392) (9/11 attack)

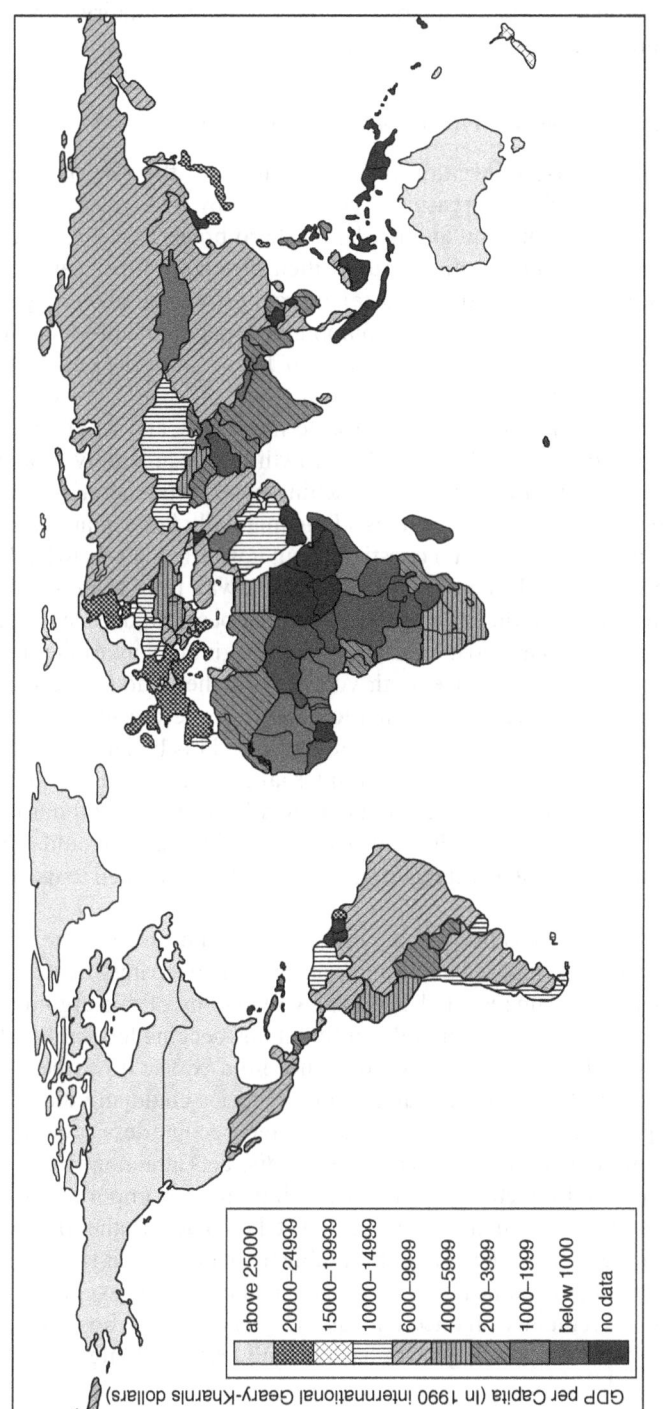

Figure 10.1 World map of GDP per capita, 2008 (1990 international dollars).

Wealthy, developed regions are the most peaceful (oil producers are the partial exception), while the poorest, least developed regions have the greatest potential for war.

The legend reads:

GDP per Capita (in 1990 international Geary-Khamis dollars)

- above 25000
- 20000–24999
- 15000–19999
- 10000–14999
- 6000–9999
- 4000–5999
- 2000–3999
- 1000–1999
- below 1000
- no data

We now proceed to examine some of the major elements involved in the Modernization Peace, in their mutual interaction. We extend our inquiry into the past, since 1815, while also tentatively peering into the future.

Industrialization or development: Escape from Malthus

An exponential increase of wealth has been central to the rise of industrial-technological society. This increase has been fueled by a steep and continuous growth in per capita production and marked a sharp break from the Malthusian trap that characterized human history until then. Production per capita for the first time registering substantial and sustained real growth at an average annual rate of 1.5–2.0 per cent. As these rates compounded, per capita production in the developed countries has increased by a factor of 15–30 since the outbreak of the Industrial Revolution.[34]

Hence the first major change with respect to modernity and its effect on war: the pie has been growing steadily, and this growth has been largely grounded in internal development, in the adoption of machine-based production, the unbound Prometheus. Wealth no longer constitutes a fundamentally finite quantity as it was in premodern times, when the main question about it was how it was to be divided. The acquisition of wealth has progressively shifted away from a zero-sum game. Auguste Comte expressed the growing feeling in the peaceful decades after 1815 when conceptualizing that warrior society had been giving way to the industrial stage of human development.[35] Helmuth von Moltke, the famous future chief of the Prussian general staff, voiced a similar logic in 1841, before autocratic Prussia had either a parliament or a constitution. He expressed his belief in the idea of a lasting peace in Europe, pointing out that internal growth in Prussia since 1815 had exceeded any advantage that could be gained by war, and without its costs and sacrifices.[36] Indeed, Prussia, the proverbial militarized state, would remain in peace for 49 years, until 1864, a staggeringly long period compared to its pre-1815 record.

It is not that conquest has become unprofitable in and of itself in the industrial and information ages as opposed to the agrarian age, when territory mattered most, as some scholars have suggested. There are two versions of this argument: that war has become inherently unprofitable, or that it has become less profitable than earlier in history. The first proposition does not hold water. It has been shown that the occupation of developed countries or territories could and did pay off.[37] The other proposition, that war and conquest have become less profitable than before, is difficult to quantify. But, in any case, it distracts attention from the main point. While cost and benefit constitute a continuum, it is important to realize where exactly on that continuum the change has taken place. Rather than war and conquest becoming less profitable or unprofitable in themselves, it is peaceful economic activity that has become a more profitable and promising avenue to wealth, while conquest has remained an uncertain and risky endeavour whose diversion of resources away from economic growth has become less appealing. The question is not whether a system of conquest and military possession pays off, but whether the alternative of production (and trade) pays more. It is peace that has become more profitable under modern conditions, tilting the relative balance between the two behavioural strategies.[38]

Indeed, in the industrial age, belligerency has decreased also among countries that have not yet embarked on the road to modernization, and for three main reasons. First, such countries have grown weaker relative to developed countries, and so have their chances of military success. The Chinggis Khan option had died out. Second, for this reason, as well as for the economic rewards, less developed countries have been incentivized to spare their resources and invest them in modernization. Third, they have become more susceptible to pressures from the hegemonic developed world, which weighs on them not to disrupt the peace.[39] Thus, economic growth in preference to war has become more attractive for developed countries, while also swaying undeveloped ones. This rationale tallies with the observed distribution of the Modernization Peace. It is far stronger in the developed world. Whereas in the past the wars among the rich and mighty were the most frequent and climactic, the great powers and other developed countries increasingly ceased to fight each other, with the developed parts of the world turning into a 'zone of peace'. The wars that remain take place among less developed countries or between them and developed countries, in the world's less developed 'zone of war'.[40] Joshua Goldstein points out the effect of international peacekeeping forces in the developing world since 1945. But, again, it should be remembered that most of the decline of war during that period has occurred in the developed world, where no peacekeeping forces exist.[41] Given this polarity, aggregating the record of belligerency of both developed and undeveloped countries compares apples with pears, as it makes no distinction between modern and largely premodern conditions (Figure 10.2).[42] That said, the Modernization Peace is also noticeable, albeit to a lesser degree, in the developing world, for the reasons explained.

At present, the logic of development and peace extends to China and the countries of East, Southeast, and South Asia, but also to aspiring countries in other, undeveloped parts of the world that have yet to succeed in getting the process going in earnest. Sub-Saharan Africa, one of the world's least developed areas, has registered steep growth rates in recent years. The marked decline in belligerency in Asia and Africa since 1990 is somewhat reminiscent of the decrease in war in post-1815 Europe. The Arab Middle East, which so far has failed to kick-start economic (and other aspects of) modernization, has been one of the world's regions most afflicted by inter- and intra-state conflict. Russia, a failing industrial developer whose economy relies heavily on the extraction of oil and gas, is a problematic case. Because more and more countries and regions are drawn into the process of economic modernization, there are good prospects for the spread of the development peace to continue and deepen.

The effect of both wealth and democracy in reducing belligerency during the entire period was first observed by Stuart Bremer. Gartzke has cited development

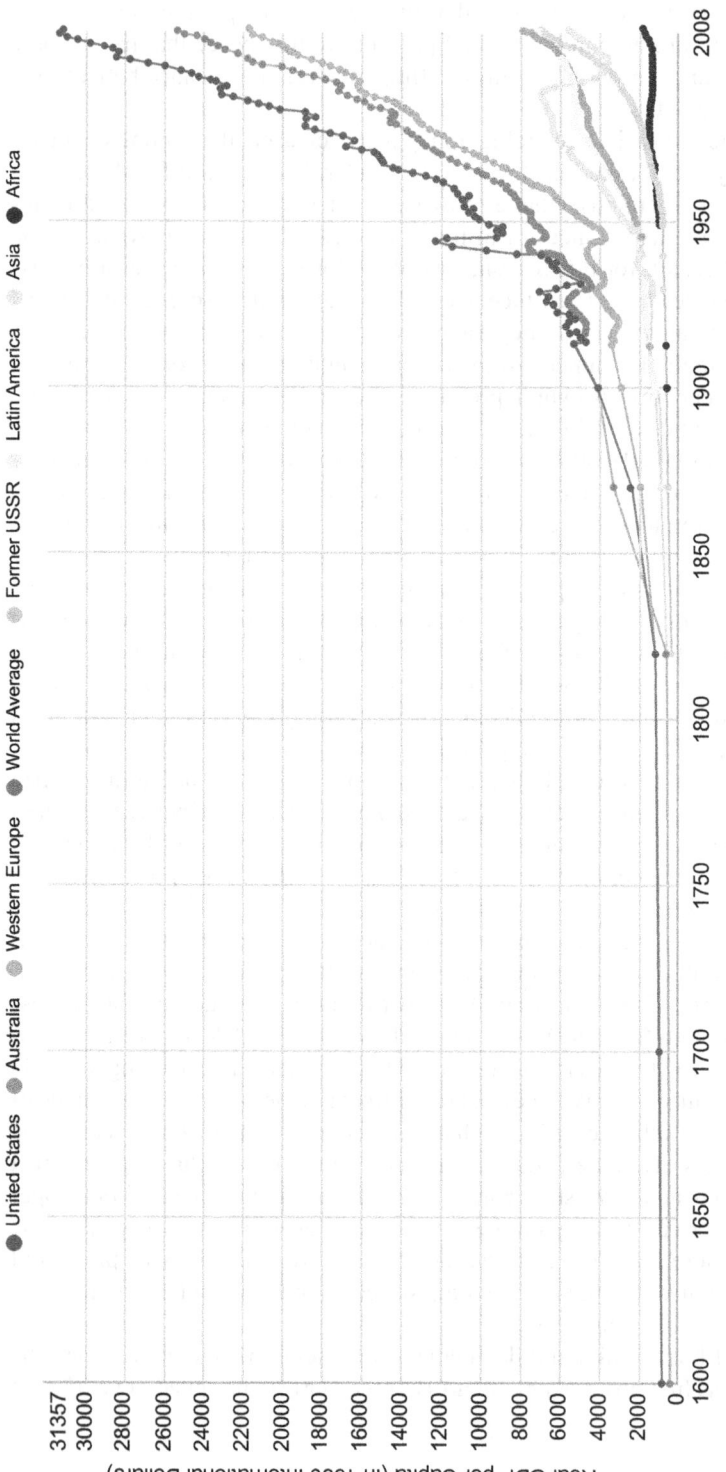

Figure 10.2 Escape from Malthus: growth in real GDP per capita (PPP).

Note the exponential rise in wealth since the beginning of industrialization, from around 1800 onwards, as compared with premodern times. Also note the still huge gap between the developed and developing countries, which makes the world average unrepresentative in many ways. However, growth in (parts of) Asia and, less dynamically, Latin America, has taken off, while Africa has yet to demonstrate that it has embarked on the road to modernization.

as a major factor in his study of the period 1950–1992. Håvard Hegre has found that it was a prerequisite of the growth of international trade during the same period. Michael Mousseau, Håvard Hegre, and John Oneal, extending their study as far back as 1885, have shown that it was also essential for the inter-democratic peace to work. The connection between development and the inter-democratic peace explains why that peace has been stronger in the twentieth as compared to the nineteenth century, and in developed as compared to undeveloped or developing countries.[43] Less noted, it explains why the inter-democratic peace was absent before modernity. Economic development accounts for the sharp decline in belligerency from the beginning of the industrial age, after 1815, among both democratic and nondemocratic, and capitalist and non-capitalist countries, albeit at different rates.

Commercialism

Commercial interdependence has been the only economic aspect of the Modernization Peace that has long gained wide recognition. The economic advantages of free trade were highlighted by Adam Smith and elaborated by David Ricardo. Still, it was not before the middle of the nineteenth century, three-quarters of a century after Smith, that Britain, the economic hegemon, abandoned protectionism and adopted free trade. Why that delay? Was it merely a matter of a good idea taking time to being recognized as such?

Although the logic of free trade was sound, it was not unlimited or unconditioned. First and foremost, its realization depended on each of the major players involved accepting the discipline of market competition as more promising for *that player* than monopolization by force. Such a reality has been tremendously boosted as the significance of trade in the economy has ballooned to entirely new dimensions with industrialization. Greater freedom of trade has become all the more attractive in the industrial age for the simple reason that the overwhelming share of fast-growing and diversifying production has now been intended for sale in the marketplace rather than for direct consumption by the family producers themselves.

During industrialization, the European powers' foreign trade increased twice as quickly as their fast-growing GDPs, so that by the advent of the twentieth century, exports plus imports grew to around half of GDP in Britain and France, more than one-third in Germany, and around one-third in Italy (and Japan).[44] Again, the greater the yields brought by competitive economic cooperation, the more counterproductive and less attractive conflict became. The various elements of this development need to be carefully unravelled.

First, there is **mutual dependence**. War between major trade partners breaks the chains of production and trade, creating shortages in the short term and inefficiency overall.

Second, there is **mutual prosperity**. Global markets mean that prosperity abroad becomes interrelated to prosperity at home, whereas foreign devastation potentially depresses the entire system and is detrimental to each country's own well-being. Thus, in stark contrast to earlier times, the enemy's economic ruin adversely affects one's own prosperity. John Stuart Mill described the radical novelty of this situation:

> Before, the patriot . . . wished all countries weak, poor, and ill-governed, but his own: he now sees in their wealth and progress a direct source of wealth and progress to his own country. It is commerce which is rapidly rendering war obsolete.[45]

What Mill formulated in principle would become an acute reality after World War I. As John Maynard Keynes warned in his *The Economic Consequences of the Peace* (1920), the crippling reparations imposed on Germany prevented its economic recovery, thereby rendering a recovery of the international economy and the resumption of prosperity among the victorious Entente powers themselves impossible. The same logic guided the reconstruction of Germany and Japan with the aid of American money in the wake of World War II.

A third variable that affects belligerency as the economy and the markets globalize is **heightened vulnerability**. As countries get involved in war, markets might be lost to others, with buyers going elsewhere to shop, foreign investment flees, and credit on the financial markets might freeze or become expensive.

Fourth, there is **open access**. An open global economic and commercial system has worked against war by disassociating economic access from the confines of political borders and sovereignty. It is not necessary to politically possess a territory in order to profit from it. Again, this enhances the logic of trade as compared to that of war and conquest. Mutual dependence, mutual prosperity, heightened vulnerability, and open access are distinct yet interconnected aspects of the commercial peace, which is, in turn, the natural complement of the industrial or development peace.

With the bitter lessons of the 1930s protectionist turn in mind, the architects of the post-1945 period in the West worked to decrease trade barriers globally. Average tariffs on manufactures were reduced from 40 to only a few per cent.[46] As during the nineteenth century, the volume of international trade in the aftermath of World War II grew twice as fast as the exploding rise in GDP, fueling the latter. With the collapse of communism and the massive advances in communication technology, globalization was further boosted. Trade in goods tripled between 1985 and 2000, while trade in capital increased sixfold.[47] As of 2012, '35 per cent of goods cross borders, up from 20 per cent in 1990. More than a third of all financial investments in the world are international transactions, and a fifth of Internet traffic is cross-border.' While the developed countries are the most connected in terms of the cross-border flow of trade, finance, and communication, the developing countries have quickly narrowed the gap, increasing their share of the world's foreign direct investments, for example, from 17 per cent in 1990 to 58 per cent in 2012.[48] This is a striking demonstration of the growing benefits that both modernized and modernizing countries derive from the system, with the potential positive effect that these benefits have on the growth of the Modernization Peace.

However, the picture is not entirely rosy or devoid of risks. As noted previously, the pacifying effects of free trade depend on each of the major players in an anarchic international system accepting the rationale of market competition as more promising than monopolization by force. Yet what happens if some of them do not? As Adam Smith himself put it, considerations of the balance of power might legitimately impose constraints on free trade, for 'defence . . . is of much more importance than opulence.'[49] Britain, the first industrial nation, experienced such a dilemma during the first half of the nineteenth century, as it embraced free trade. Thereby it boosted its own growth while fueling that of the rest of the world and connecting it to a pacifically inclined system. However, in relative rather than absolute terms, this open policy made it easier for others – above all the United States and Germany – to catch up and eventually overtake Britain economically and to challenge its position as the mightiest power. This was, and remains, a dilemma, and a gamble.

China is the most important case today. During its meteoric industrial take-off from 1979 onward that averaged an annual growth rate of nearly 10 per cent in GDP over three decades, China's foreign trade increased at double that rate. Thus, China's international trade volume as share of GDP has increased from less than 20 per cent in the early 1980s to just above 50 per cent. China's massive transformation was made possible by the favourable reaction of the United States and its liberal democratic allies to Beijing's opening to the world and embrace of the markets. They opened their markets to Chinese imports and flocked to invest in China. They did not condition this policy on political liberalization in China, believing that eventually everybody stood to gain from cheap goods, global growth, and the expansion of purchasing power; that political liberalization would in any case arrive in China in the wake of economic liberalization and development; and that China's integration into the world economy is either way the best guarantee for world peace. Whether these assumptions will materialize or, instead, a nondemocratic China will turn out to be a formidable political and military rival as its wealth and power grow remains an open question that mirrors Britain's nineteenth century dilemmas.[50]

As of 2013, China's main trading partners are, in descending order: the European Union, the United States, Japan, South Korea, and, indeed, China'sbête noire, Taiwan. All are liberal democracies and are politically and militarily allied. Similarly, China is the United States' second largest trading partner, after Canada, and the European Union's second largest trading partner after the United States. Financially, China (including Hong Kong) is on par with the United States as the largest destination of foreign direct investment, as well as being the largest holder of US treasury bonds. Thus, China's economy is symbiotically intertwined with the global economy, wherein the countries of the free world dominate. Indeed, while China accounts for 15 per cent of American trade and 13 per cent of the European Union's trade, trade with the major liberal democratic countries amounts to about half of China's trade.[51] These are facts of major significance that highlight China's vital dependency on the system, and its great vulnerability if the system were to be disrupted. Still, shadows of 1914 and the 1930s linger on.

With the onset of the Great Recession in 2008, international trade in goods, services, and finance shrank from above 50 per cent of global GDP in 2007 to just over 35 per cent in 2012, and has not recovered since.[52] While decreasing consumption and credit problems have been partly responsible for this development, the trend has also been affected by increasing protectionist pressures and currency wars between the major countries. Some observers are concerned that the second global age may have peaked. True, the growth of international trade as share of global GDP must reach a plateau at some point. Furthermore, as the gap between the developed and developing countries narrows and labour costs in the latter rise, there may be lesser incentives for the movement of production away from home. Robots and three-dimensional printing are expected to take the automation of production to a new level, further lessening the incentives for offshoring.

While the economic equilibrium point that the level of international trade finds is one thing, protectionist *political* actions that affect the openness of the system are quite another. As both the economic and political benefits of free trade are clearer today than they were either before 1914 or in the 1930s, and the harsh lessons of the two world wars are also clear, there are very strong incentives for the open trade system and commercial peace to persist. China, for example, acutely dependent on the importation of raw materials – from oil to metals to foodstuffs – needs to export in order to pay for them. Thus, if denied easy access to the world's markets, China's policy with respect to the oil resources of the South and East China seas, for example, already alarming, might become overtly aggressive. It remains to be seen if and how the Brexit vote in Britain is going to affect the freedom of trade in Europe and to what extent US President Trump's protectionist campaign slogans will be translated into policy. If so, we may cross a historical watershed. Regardless of these political developments, some observers are concerned that the second global age may have peaked.

Notably, the openness of the world economy does not depend exclusively on the democracies. China has profited enormously from the current system. Yet further down the twenty-first century, as China grows wealthier, its labour costs rise, and its current competitive edge diminishes, China itself might face the temptation to become more protectionist. As a huge market, it may choose to go on its own or in cooperation with its regional neighbours. Such reversals have occurred in the past, engulfing, not least, the paragon of free trade and the system's long-time hegemon, Britain, in 1932.

Recently, China has been creating alternative institutions to the International Monetary Fund and World Bank, which it regards as being overly dominated by the United States. In 2014, Brazil, Russia, India, China, and South Africa (the BRICS countries) signed a deal to create a development bank and emergency reserve fund. In early 2015, despite the US's outright objection, its closest allies – including Britain, Germany, France, Italy, Holland, Denmark, Australia, and South Korea – joined the Chinese-led Asian Infrastructure Investment Bank, which ultimately comprised 57 founding states. The United States and Japan were the only major economies that did not join.

Unlike China, early twenty-first century Russia is a failing modernizer. The communist failure was followed in the post-Soviet period by problematic and painful privatization, by an erosion of Russia's industrial base and by the export of natural resources in Putin's rentier state. By 2014, the share of oil, natural gas, and minerals had grown to more than two-thirds of Russia's exports, close to one-third of its GDP, and more than half of the state's revenues. The overwhelming share of Russia's trade is with the European Union, the destination of about half of Russia's exports in 2013. By comparison, China, Russia's second largest trade partner, consumed only 7 per cent of Russia's exports, while Russia ranked only in ninth to 10th place among China's trading partners. Russian-American trade is negligible. In terms of the volume of trade, Europe's dependency on trade with Russia is lesser than Russia's dependency on trade with Europe. In 2013, Russia was the European Union's third largest trading partner, and trade with Russia comprised less than a tenth of the EU's trade. However, importing close to 40 per cent of its gas and nearly a third of its oil from Russia, Europe's dependency on Russia is as critical as Russia's dependency on trade with Europe.[53]

In 2014, Russia's occupation and annexation of Crimea and military intervention in the Ukraine sparked the most severe international crisis between the powers since the end of the Cold War. The United States and the European Union have adopted a policy of limited economic sanctions, hoping to tame Russia without escalating the crisis. Europe's dependency on energy imports from Russia has in any case ruled out the possibility of harsher sanctions. Nobody can tell how events will unfold, and this article is in any case not about predicting the future. The Ukrainian crisis serves here merely as an illustration of how nationalist sentiments on the one hand and the rationale of economic growth and trade dependency on the other may conflict, as they have in the past, heightening security concerns, accentuating the security dilemma, and threatening the Modernization Peace. We shall elaborate on this point later on.

Furthermore, while the reasons for Russia's military action in the Ukraine are multifarious, economic considerations are not absent among them. To all appearances, Russia is a lesser participant in the development peace and a greater one in the commercial peace due to its energy and minerals exports. However, in reality, Russia's condition as a failing modernizer extends from production to trade. Because of the failure of its industrial sector (except for the military industry) to compete in the world markets, Russia is inclined to resort to political pressure in its trade relations with close neighbours and previous members of the Soviet imperial system, who have their eyes fixed on the much more attractive economic connection with the European Union.[54] Thus, at present, Russia does not view the open trade system as advantageous as far as its own manufacturing sector is concerned. Indeed, Russia views it as a threat. Having failed to integrate into the global system of production, trade and mutual affluence in the 1990s, Russia has retreated to a bullying state system both at home and abroad. Russia's imperial traditions with respect to its small neighbours and historical satellites are in any case strong. Former US Secretary of State John Kerry condemned the Russian invasion of the Ukraine, stating that 'you just don't in the 21st century behave in 19th century fashion'. However, what the twenty-first century will be like remains to be seen.

Affluence and comfort

The industrial-commercialized economy has spawned another new and distinct factor over time that has contributed to the decrease in belligerency: affluence. Although the concept of affluence has become central to the analysis of contemporary society, its major effect on war has attracted little attention. Throughout history, rising prosperity has been associated with a decreasing willingness to endure the hardship of war. That luxury breeds softness was an age-old philosophical-moralist maxim in all the classical civilizations. As the industrial-commercial age unfolded and wealth per capita rose exponentially, the wealth, comfort, and other amenities once enjoyed only by the privileged elite have spread throughout society. Affluence – the freedom from scarcity in the elementary, vital needs of life and the ability to afford a great variety of consumer goods – has been achieved throughout the societies of the developed world in the wake of World War II, when the concept itself was coined. Affluence levels have continued to rise ever since. Thus, affluence has been a latter-day add-on to the Modernization Peace, contributing to and much enhancing it in the period after 1945.[55]

Some familiar social science concepts are applicable in elucidating the effects of affluence on decreasing belligerency. People embrace 'postmodern', 'post-materialistic', hedonistic values that emphasize individual self-fulfillment. As Ronald Inglehart has documented, people have been moving away from the 'survival values' of old to the 'self-expression values' characteristic of affluent societies.[56] Another relevant concept is Norbert Elias's 'civilizing process'.[57] In an orderly and comfortable society, rough conduct in social dealings decreases, while civility, peaceful argument, and humour become the norm. Yet another familiar and highly relevant concept is 'risk aversion'. Once the vital necessities of life are secured, people tend to adopt more conservative, less desperate, and less risky behavioural strategies. On a lighter note, New York Times journalist Thomas Friedman suggested in the 1990s that a McDonald's Peace prevailed between any two countries that had become rich enough to have the popular restaurant chain. In the meantime, there have been exceptions to his rule, but the basic idea stands.

Indeed, more so than the development and commercial peace, the affluence peace is absent in the least developed parts of the world, contributing to the relative weakness of the Modernization Peace there. As we have seen, the top 30 per cent most affluent countries – roughly above 13,500 GDP per capita in 2013 USD – do not fight among themselves, nor experience intra-state armed conflict (with oil countries again somewhat diverging from the general trend). Returning to the most vital case with respect to the future of the international system, China's wealth per capita as of 2015 is around 7,000 USD (in rate of exchange terms), one-fifth to one-sixth that of the developed countries (or one-fourth to one-fifth in purchasing power parity). Its affluence rates are more or less in the same range as those of the most advanced societies during the world wars era, which exhibited considerable enthusiasm for the war in 1914 and a marked lack of enthusiasm 25 years later. As China's wealth doubles, supposedly by the late 2020s, and its comfort levels rise, the Chinese people's willingness to endure the risks and hardship of war (not high today) may enter much safer territory. On the other hand, China's power, international clout, and potentially more assertive political claims may increase in proportion to its growing wealth. Thus, conflicting processes and effects may be at work here, and it remains to be seen how they will unfold and interact.

Urbanism and urbanity

The 'civilizing process' is intimately associated with one of modernity's major developments: the process of urbanization and the growth of metropolitan-service society. Commercial and metropolitan cities were considered by classical military authorities, such as Vegetius, echoed by Machiavelli, as the least desirable recruiting ground, compared to the countryside with its sturdy farmers accustomed to hard physical labour. With modernity, urbanism in large metropolises steadily expanded to encompass the majority of the people. Correspondingly, the number of people living in the countryside plummeted. Nonetheless, the military in the major countries of the developed world during the twentieth century continued to regard them as the best 'recruiting material'.[58]

Moreover, city folk during the zenith of the industrial age were mainly factory workers. They were accustomed to physical labour, machines, and the massive, coordinated work regime labelled 'Fordism' or 'Taylorism'. They lived in dense urban communities and were mostly literate. These qualities were major strengths for the military, especially as the military, too, was undergoing mechanization. However, as the industrial-technological era progressed, manufacturing declined and the services sector expanded its share of the workforce in the most advanced economies. In the United States, which led this trend, 70 per cent of the workforce is now employed in services while only 18 per cent work in manufacturing.[59] It can be argued that the military, too, has been moving from mechanized to information-based forces, increasingly relying on computerized data processing and accurate stand-off fire to do most of the fighting. All the same, adaptation to military life comes far less naturally to people from contemporary affluent societies who are accustomed to deskwork in the office and the seclusion of residential suburbia than it did to their farmhand and factory-worker predecessors. Again, the world's least developed countries are less affected by this development.

The world in general has recently crossed the 50 per cent urbanism rate, an average that bridges over a broad spectrum between the developed and developing countries. In the former, two-thirds of the population or more live in urban centres.[60] China crossed the 50 per cent point in the 2000s, as Germany had done around 1900, and Britain around 1850. Wealth per capita in China's industrial cities is 2–3 times the national average, and indulgence in the good life is very noticeable there. As of 2008–2010, China's labour force was divided almost equally between the agricultural, industrial, and service sectors. By comparison, in most developed countries the service sector accounted for close to four-fifths of the economy and the agricultural sector comprised only a few per cent.[61] Over time, China's urban-urbane features are likely to deepen.

Liberalism and democracy

Let us repeat what we know about the (inter-)democratic peace after several decades of intensive debate. It is a very real phenomenon. It sets liberal democracies sharply apart from nondemocratic countries, which are involved in wars both with democracies and among themselves (although they too have fought less, and not only recently, as is increasingly recognized, but ever since 1815[62]). It has grown stronger over time, becoming more entrenched during the past century than it was during the nineteenth century (and, less recognized, there is no trace of it earlier in history). It is stronger among economically developed than among developing countries.

How do all these different features fit together and how can they be coherently explained? The continuous deepening of the democratic peace and its greater strength in the developed countries have both been attributed to the deepening of democracy itself and of liberal values over time.[63] However, is this a purely cultural-moral-social attitude development that feeds on itself? If so, why do economically less developed societies lag behind in this process?

The connection between socio-economic modernization and the growth of liberalism and democracy has long been noticed.[64] True, there are some poor democracies, most notably India since 1947. But there, too, democracy would have been unimaginable without networks of modern communications: railways and automobiles, newspapers, telegraph, radio, and television. On the other hand, there is also the question of developed nondemocratic countries, discussed below. All in all, however, the record suggests that democracy becomes more entrenched with economic development, and that there is an interconnection between the two factors and the growth of the democratic peace.

Furthermore, although nondemocratic countries in the industrial era have also participated in the general decline in belligerency, liberal and democratic societies have proven most attuned to modernity's pacifying aspects. Relying on arbitrary coercive force at home, nonliberal and nondemocratic countries have found it more natural to use force abroad. By contrast, liberal democratic societies are socialized to peaceful, law-mediated relations at home, and their citizens have grown to expect that the same norms be applied internationally. Living in increasingly tolerant societies, they have grown more receptive to the Other's point of view. Domestically promoting freedom, legal equality, and political participation, liberal democratic powers – though initially in possession of vast empires – have found it increasingly difficult to justify rule over foreign peoples without their consent. And, sanctifying life, liberty, and human rights, they have proven to be failures in forceful repression. Furthermore, with the individual's life and pursuit of happiness elevated above group values, sacrifice of life in war has increasingly lost legitimacy in liberal democratic societies. War retains legitimacy only under narrow and narrowing formal and practical conditions, and is generally viewed as extremely abhorrent and undesirable.

Democratic leaders have either shared these values, or have been forced by public pressure to conform to them, or have been removed from office. The last option, the so-called 'structural' explanation for the democratic peace, makes no sense on its own. It only makes sense *after* a change in the popular preferences coupled with a 'normative' change in the attitude towards war.[65] Unless the perceived utility and justification of war are generally tilted in the negative direction in the eyes of the electorate, voters should be presumed to reward victorious leaders as much as punish defeated ones.

With growing liberalization, democracy, and economic development, the probability of war between democracies has declined to a vanishing point, with the 'security dilemma' itself remarkably disappearing in their relations with each other. Domestically too, on account of their stronger consensual nature, plurality, tolerance, and indeed, a greater legitimacy for peaceful secession, developed liberal democracies have become practically free of civil wars, the most lethal and destructive type of war.[66]

The forceful democratization of Germany and Japan after World War II, the most significant cases of democratization in the twentieth century, was made possible not only by the political circumstances of defeat. Both countries possessed a modern economic and social infrastructure upon which functioning liberal democracies could be built.[67] The Arab Middle East offers a sharp contrast to Germany and Japan, and a vital clue as to the close connection between modernization, democracy, and peace. This applies equally to the American campaign to bring democracy to the Arab world and to the locally induced Arab Spring that has collapsed into a bloody mayhem.

There is a catch that lies at the root of affluent liberal democracies' torment in conflict situations. Since wars are abhorred in liberal societies as antithetical to both their interests and values, to their entire way of life and worldview, they are sanctioned only as a last resort – after all other options have failed. Yet in practically no situation does it ever become clear that all alternative policies have indeed been exhausted, that war has really become unavoidable. A feeling that there may be another way, that there *must* be another way, always lingers on. Errors of omission or commission are ever suspected as being the cause of undesired belligerency. Moreover, it never becomes clear that the democracies come to a conflict with entirely clean hands, morally, because of past or more immediate alleged wrongs; nor indeed can they, given the inevitable gap that always separates ideals from reality.

In conclusion, liberal democracy has had a very distinctive effect in reducing war during the past two centuries, but only in connection with the economic development that has revolutionized the world since the advent of the industrial age. We shall discuss the salient question of democracy in China and Russia later on.

Sexual liberalization – make love, not war

The revolution in sexual mores and the advent of sexual promiscuity is another factor that has dampened enthusiasm for war in advanced modern societies. Throughout history, rape, like looting, was one of the major perks of war. The troops of Imperial Japan and the Soviet Union still practised mass rape during World War II, in the Japanese army also in the form of state-organized forced prostitution. Mass rape has been a major feature of the recent wars in Bosnia, Rwanda, Darfur, and West Africa. In the armies of the Western democracies, American and other Allied troops widely availed themselves of an abundant supply of low-cost prostitution in ruined Western Europe after World War II and, later, in desperately poor Vietnam.[68]

All in all, however, the much increased sexual opportunity within society sharply decreased the incentives to enlist for war. Young men now are more reluctant to leave behind the pleasures of life for the rigours and chastity of the battlefield. 'Make love, not war' was the slogan of the powerful anti-war youth campaign of the 1960s, which not accidentally coincided with a far-reaching liberalization of sexual norms. Barely counted at the level of national or foreign policy aims, the sexual revolution deeply affects national policy through the bulk of individual choice.

Expanding sexual freedom is part of the 'good life' indulged in by much of the world today, and it constitutes an important element of the Modernization Peace. It is prevalent in both China and Russia. Sub-Saharan Africa displays both sides of the coin. Societies there generally exhibit quite relaxed sexual mores, but in quite a few of them there are also incentives to join sectarian, tribal, and private armies that regularly engage in mass rape. In the ultra-conservative Arab world, highly restrictive sexual norms are a major factor driving young men's belligerency. Indeed, virgins are on offer to jihadists not only in paradise, but sadly also in this world.

Aging demographics

In addition to changes in the circumstances and attitudes of young males, the significant decline in their relative number is another factor that may have diminished enthusiasm for war in contemporary developed societies.[69] With the onset of industrialization, a youth bulge emerged in the nineteenth century West, as it did in the twentieth century developing world. Young male adults were most conspicuous in the public enthusiasm for war in July-August 1914, as they were in all wars and revolutions. However, in today's world, with birth rates falling and with increased longevity, young adults constitute a shrinking portion of an aging population. Before World War I, males aged 15–29 constituted 35 per cent of the adult male population in Britain, and 40 per cent in Germany; by the advent of the twenty-first century, their share had dropped to 24 per cent and 29 per cent, respectively.[70] In many developed countries, such as Japan, South Korea, Taiwan, Italy, Spain, and Germany, but also in less developed or developing countries such as China and Russia, fertility has fallen far below replacement rate, and the population is quickly aging. Because young males have always been the most aggressive element in society while older men were traditionally associated with counsels of moderation and compromise, the youth bulge may have increased the enthusiasm for war in the earlier stages of modernization, while an aging population contributes to its decrease at a later stage.

The relationship between young men's sexual opportunities, the youth bulge, and the demographic peace is of particular interest with respect to Iran. Iran's

GDP per capita was just over 6,000 USD in 2013 (91st out of 193 countries), much of it derived from oil. Iran's urbanization rate is around 70 per cent. Among the urban more affluent classes, indulgence in the good life, liberalizing norms, opposition to the Islamist regime, and aversion to the prospect of war are increasingly noticeable. However, all of those factors apply much less to the peasants and urban poor, who together comprise the great majority of the population. The demographic transformation of Iran is far more sweeping, and promising. The fertility rate in Iran has plummeted from 6–7 children per woman in the 1980s to less than 2. In consequence, the share of young males aged 15–29 in the population, which constituted 48 per cent of all males in Iran in 2000, is fast dropping and is expected to halve by 2050 and become lower than in Britain's and almost as low as in Germany. This dramatic change, like other changes in Iranian society, could have many effects over time, including a decrease in belligerency.[71]

Fertility rates in some Arab countries have also been dropping sharply (though less radically than in Iran), while in others, as well as in Islamic countries of Central Asia, including Afghanistan and Pakistan, the decrease has been much slower. This means that these parts of the world still have a huge youth bulge, strongly contributing to their deep deficit in everything relating to the Modernization Peace. At the same time, long-term prospects are improving, at least by this measurement. Fertility rates in Sub-Saharan Africa, although also decreasing, are still by far the world's highest, with a corresponding youth bulge.[72]

Whatever other problems this trend brings, the pacifying aspects of an older social age structure and something like a youth deficit seem to be quite straightforward over time.

Women's political participation

While young men have always been the most aggressive element in society, men in general have always been more aggressive and belligerent than women. Having won the right to vote in twentieth century liberal democracies, women have been able to influence governments' policies. Studies have shown a consistent gender gap in attitudes toward the use of military force.[73] Such differences might play a significant role in tilting the electoral balance against military ventures in modern affluent liberal democracies.[74]

Intriguingly, the world's two most populous state societies send warning signals with respect to the gender element of the Modernization Peace. Because of a state-enforced one-child policy and preference for a male child, the people of China have been practising female abortion and infanticide. As a result, China holds the world's most extreme sex ratio, with 116 boys to 100 girls under the age of 14. Similar preferences and practices in India make that country almost as extreme a case, with 113 boys to 100 girls under the age of 14.[75] The female deficit also means that sexual prospects for many young males are bleak. How much weight this factor will carry within the plurality of elements that comprise the Modernization Peace is difficult to say.

Cultural convergence and divergence

The globalization of mass culture and cultural convergence are such widely discussed elements of modernization that we find it unnecessary to expand on them. Shared cultural themes disseminated by the media, the entertainment industry, and more recently by the internet and social media, decrease the difference in outlook and values between societies, and make them appear more familiar and less alien to one another. The limitations of this powerful trend have also been recognized. Local cultures remain deeply entrenched and, indeed, may sometimes react to the pressures from 'global culture' with self-assertiveness, hostility, and even violence. Those who reject modernity or champion it in forms different from the liberal Western model have reacted most strongly, and have, indeed, been using the internet and social media most effectively to spread their message. While the globalization of trade and finance has stalled since the outbreak of the Great Recession in 2008, the flow of information across borders via the internet has continued to grow. At the same time, the internet has also spurred a massive expansion of indigenous cultural expression.[76]

Past and future challenges to the Modernization Peace

As previously noted, a comprehensive and coherent understanding of the Modernization Peace needs to account for both the general decrease in belligerency since 1815 and the monumental exception to this trend: the two world wars. Thus, we begin with past divergences from the Modernization Peace, in order to both explain major lingering puzzles and enrich our overall understanding of the forces concerned, not least in order to see what these may augur for the future. While anything resembling rigorous prediction is not possible, an attempt to identify some of the fundamental factors involved in the modern transformation and affecting the Modernization Peace – including both their scope and limitations, as well as of counter-processes, actual and potential – is crucial for both retrospect and prospect, and at both the theoretical and practical levels.

Why the great wars of the nineteenth and twentieth centuries?

Strugglers in the process of modernization have continued to engage in wars, as they still do. However, wars during the nineteenth and first half of the twentieth centuries also broke out between countries that were among the leaders along the path of modernization. How can this be explained?

We begin with the great power wars that took place between 1854 and 1871, between the First and Second Long Peace. These were the Crimean War (1854–1856), the Franco-Austrian War (1859), which led to Italy's unification, the American Civil War (1861–1865), and the Wars of German Unification (1864, 1866, 1870–1871). With the exception of the Crimean War, caused by traditional great power strategic considerations, it was above all issues of national unity, national independence, and national self-determination that constituted the deepest and most inflammable motives for these major wars. Whereas territory is documented to have been the main motive for war in early modern Europe, with trade as the second most common motive, ethno-national conflict is recorded to have been by far the primary cause of major war in the nineteenth century.[77] Thus, while the logic of the new economic realities worked against traditional material causes of war, that logic was sometimes overridden by the rising tide of modern nationalism.

During the past decades, ethno-national wars within the most developed parts of the world have all but disappeared. This is partly due to growing democracy, liberalism, and respect for minority rights, which helps to accommodate ethnic minority groups. However, the decline owes no less to the growing acceptance of the right for national self-determination voted for by an aspiring national group as a legitimate justification for secession. Quebec, Scotland, Belgium, and Catalonia are some major recent cases in point. This development is in itself part of the Modernization Peace. Indeed, it has far less affected the less developed parts of the world, wherein political and ethno-national boundaries often do not overlap, while peaceful secession is seldom allowed. In Africa and Asia in particular, many countries have a complex ethnic mosaic, and quite a few of them are beset by ethnic conflicts, with potentially more in store. Thus, ethno-national conflicts constitute the main cause of war in the developing parts of the world, as they did in nineteenth century Europe.[78]

Nationalism was also a major contributing factor in the return of great power war during the first half of the twentieth century. By 1900, the great powers, except Britain, had resumed protectionist policies and were expanding them to the undeveloped parts of the world with the New Imperialism. The fear that the emergent global economy might become closed rather than open, as the world was being carved into large imperial blocs, grew in the system. Each power felt obligated to snatch up as many of the territories that were up for grabs as it could, while it could.[79] It was this 'prisoner's dilemma' situation that drove the powers to press their claims in Africa and China. The British-dominated free trade system was being eroded by the reality and prospect of protectionism, which was a self-reinforcing process and self-fulfilling prophecy. The new industrial giant Wilhelmine Germany in particular felt that it had to escape the constraints of its narrow political-economic boundaries within Europe and expand into a world power. Here was the main driving force behind the rising tensions and repeated crises among the powers in the decade before World War I, which eventually erupted into war.

The slide towards national economic protectionism accelerated with the Great Depression. In 1932, Britain abandoned free trade and adopted 'Imperial Preference'. The United States, which itself raised tariffs in the wake of the 1929 crash, would increasingly view the growth of protectionism as inhibiting its recovery. However, by far the most ominous consequences concerned Germany and Japan. For Adolph Hitler, the creation of an economically self-sufficient German Reich whose *Lebensraum* would bestride continental Europe was inseparable from his racist plans and vision of a perpetual global struggle. Similarly, as Japan found itself barred from international trade by the rise of protectionism, it regarded the establishment of its own economically self-sufficient empire, or 'Greater

East Asia Co-Prosperity Sphere', as essential to its survival. Correspondingly, Japan's liberal parliamentary regime during the 1920s gave way to militarism and authoritarianism.

Thus, major interconnected elements of the Modernization Peace were replaced by their antitheses. In a partitioned global economy, economic power increased national strength, while national strength defended and increased economic power. National size made little difference in an open international economy, but became the key to economic success in a closed, neo-mercantilist international economy dominated by power politics. It was again becoming necessary to politically own a territory in order to profit from it. Furthermore, the retreat from economic liberalism spurred, and in turn was spurred by, the rise to power of anti-liberal and anti-democratic political ideologies and regimes, incorporating a creed of violence: communism and fascism.

Is alternative modernity viable?

The question that these past experiences raise concerns the viability of alternative forms of modernization – different from the liberal democratic and capitalist – and their affinity with the Modernization Peace. The liberal political and economic order established in the nineteenth century by Britain has been variably challenged by later industrializers and newcomers to the modernization process.

Lowest on the scale of challenges, the economic protectionism increasingly adopted by liberal democracies between 1870 and 1945 did not noticeably increase their belligerency vis-à-vis one another. As we have seen, the pacifying economic elements of industrialization encompass more than free trade, and these elements, together with the other – political, social, and normative – elements of the Modernization Peace held, indeed, continued to grow, in the democracies during that period. On the other hand, protectionism did increase the propensity of liberal democracies to get involved in (largely reactive) wars with industrializing non-democratic/non-liberal countries. Although the latter also exhibited a very marked decrease in belligerency compared to premodern times, their residual tendency to fight among themselves and against liberal democracies escalated when strong protectionist pressures pervaded the system and reinforced nationalistic tensions. Further up the scale, fascism (as a broad generic term) constituted the most sweeping rejection of all aspects of the liberal political and economic model, and its strong militant streak shattered the Modernization Peace. Finally, communism was another projected alternative route to modernity. While the Soviet bloc shared in some aspects of the Modernization Peace – such as the premium given to internal industrial growth – sharp ideological differences sustained by economic self-sufficiency involved the Soviets in an intense Cold War with the non-communist world both before and after World War II.

The supreme relevance of this question derives from the return to the international arena of capitalist nondemocratic great powers, above all China, but also Russia (see chapter 8 in this volume). What effect China's rise will have on the international system, American supremacy, the liberal hegemony, and the

Modernization Peace may be the most significant question of the twenty-first century. Furthermore, even if the peaceful scenarios for China's continued rise eventually materialize, what major convulsions, including militarized confrontations, may shake the world before the process is completed? All these are much discussed questions that at present can only be regarded as thought experiments.

With the outbreak of the Great Recession in 2008, the world has experienced the worst economic crisis since the Great Depression. Indeed, analogies to the 1930s, when fascist and communist totalitarianism throve on the apparent failure of capitalist democracy, are inescapable. One hopes that the current economic crisis will not be nearly as catastrophic politically. And yet the hegemonic model's loss of much of its aura, as in the dramatic reversal in the image of the European Union, from the paradigm of the future to a deeply problematic and dysfunctional basket case, has left a strong impression. The recent election of Donald Trump as president of the United States has shocked many, and only time will tell to what extent their concerns are justified. The more dysfunctional and crisis-ridden the liberal democratic countries appear, the greater the self-confidence and global allure of state-driven and nationalist capitalist authoritarianism. The Third Wave of democratic expansion since the 1970s has stalled from 2006 onward, and commentators are increasingly discussing an 'authoritarian resurgence'.[80] Latin America, Central Asia, the Middle East, and Africa may be particularly susceptible to the capitalist nondemocratic model.

By the same token, the massive challenges and obstacles that China faces and its structural weaknesses, as well as huge and yet unfulfilled potential, are both quite evident. The ability of China's regime to fight corruption, retain domestic legitimacy when setbacks and crises inevitably occur after a long period of economic growth, and withstand pressures to embrace political liberalization and democratization remain open questions. Internationally, the more assertive China becomes vis-à-vis its neighbours, the more it might generate a backlash and actually lose influence over them. Indeed, in a replay of Germany's tragic course a century earlier, China might prompt a coalition to contain it, stretching from Japan to India, with the backing of the United States.

Russia is in an entirely different league than China. It is 10 times smaller in terms of population and its oligarchic and kleptocratic brand of authoritarian capitalism is shakily built on the 2000s bonanza in the price of oil and gas. Despite its renewed assertiveness, Russia remains a poor and, on the whole, weak country, and is unlikely to break through to the rank of the advanced economies unless it is able to revive its manufacturing sector, building on its educated workforce. Russia has a long tradition of deep ambivalence towards Western values, stretching as far back as the early modern period, if not earlier. Envy of and a sense of inferiority towards the Western model have coexisted and alternated with a strong sentiment of Russian and Slavic special identity and mission: religious, spiritual, and ideological. Perception of the West as advanced, civilized, and successful has always competed with its image as immoral, profane, and degenerate, celebrating individual egotism and hedonism, and practising false morals and double standards. There are good reasons to think that Russia's deep historical ambivalence towards all things Western, resurrected under Putin, will continue into the future.

I do not profess to know whether or not a capitalist nondemocratic alternative modernity is here to stay. Political liberalization and democratization in China, if they occur (and if they go hand in hand), may become the final major step towards a global ascendency of the Modernization Peace. However, while liberal democracy is a strong contributing factor to the Modernization Peace, this peace is a broad phenomenon that also encompasses nondemocratic (and non-capitalist) countries which share elements of it other than democracy. Thus, peace may prevail even if China and Russia do not democratize, or takes very long to do so. Nobody knows which of many possible paths the future will take.[81] It is a coin toss. The entire spectrum is possible: from more or less cordial economic and political cooperation, to heightened ideological rivalry and arms races fueled by the security dilemma, to cold war and limited wars – with fluctuations between them all. In both China and Russia the regime's insecurity and rising nationalism are a potentially combustive blend. Ultimately, only nuclear deterrence may prevent a repeat of what Bismarck called 'some damn foolish thing in the Balkans' from materializing in the South or East China Seas, around Taiwan, or in the Baltics, and setting off a major war.

Failed modernizers and unconventional terror

Neither terrorism nor weapons of mass destruction are new. Similarly, failed modernizers and anti-modernizers – the two being often connected – are as old as the process of modernization itself. It is the joining of these elements, the ability of small groups or even individuals – often motivated by fanatic anti-modernist ideology and willing to die in pursuit of their cause – to resort to weapons of mass destruction, that creates the new potential of the present and future.

Critics of the 'War on Terror' after 9/11, most notably John Mueller, argue that the threat is massively overrated.[82] However, while demonstrating the limited potential of a chemical terror attack and the enormous difficulties surrounding nuclear terror, Mueller concedes that bioterror 'could, indeed, if thus far only in theory, kill hundreds of thousands, perhaps even millions of people'.[83] Indeed, the biotechnological revolution of our times has made biological weapons both much more lethal and more accessible, and the trend is likely to continue. Not only might a successful biological attack result in casualties on par with the United States' greatest wars; it is likely to target the country's main population and economic centres. One would be ill-advised to make light of this threat.

The root of the problem lies in the technologies and materials of WMD trickling down to below the state level. The encapsulation of destructive power creates a situation in which a player no longer has to be big in order to deliver a devastating punch. Hence the alarm raised as jihadist Islamist organizations win control over large territories. Recent years have seen the collapse of state authority in a number of countries in the developing world, most notably Syria and Iraq, Yemen, Libya, Somalia and north Nigeria. In all of them, militant Islamist organizations thrive. The rapid expansion of the Islamic State of Iraq and Syria (ISIS) in 2014–2015, coupled with the organization's virulent anti-modernist ideology and hideous practices, have attracted particular attention.

The Arab and Muslim societies from which such groups arise are generally poor and weak. Only the potential use of WMD makes the threat of militant Islam significant. Furthermore, other causes and 'super-empowered angry men' (as Thomas Friedman has called them) would always be present and could now assert themselves with horrific consequences. While societies in general tend to become more pacific as they modernize, there will always be individuals or small groups who will embrace massive violence for some cause.

Contrary to the classical argument regarding the pros and cons of nuclear proliferation, the greatest threat of nuclear and other forms of nonconventional weapons proliferation may lie in the increased danger of leakage or anarchy in the less developed parts of the world. Nuclear Pakistan demonstrates both dangers. The collapsed Soviet Union of the 1990s and the concerns regarding its nuclear arsenal and unemployed scientists, rather than the former nuclear superpower, may be the model for future threats to the modern peace.

Conclusion

The processes of modernization that underlie the spread of peace have been real and deep. Indeed, they have been among the most sweeping in human history, and their effect has been nothing short of revolutionary. The debate on whether democracy or trade/capitalism have been responsible for the decrease in war is somewhat off the mark, as both have been preconditioned by a broader and more fundamental factor: modernization, sparked by the process of industrialization from 1815 onward. The legs of Kant's tripod for peace are not only connected but must also stand on something: they are firmly grounded in the industrial age.

Industrialization brought about an exponential increase in the role and weight of trade, which together with rocketing wealth, rooted in internal growth, have tilted the preferences of people and states away from war and towards peace. This change was most acutely, though not exclusively, manifested in the choices made by democratic electorates. Other offshoots of the modernization process have deepened the trend. The development, commercial, affluent, urban-urbane, liberal democratic, sexual liberation, demographic, gender, and cultural convergence factors are all interconnected and mutually affecting in the modern transformation and in the growth of the Modernization Peace. At the same time, they each exercise a distinct contributing effect that comes into play at different points in time, reinforcing the war aversion of the societies involved. Other factors within the modernization process can surely be defined. Taken together, the above factors help to account for the gaps and reconcile the discrepancies that close historical control reveals in theories such as the nuclear peace, the democratic peace, and the capitalist peace, now understood within a more comprehensive and unifying whole. There is no Ockham's razor here by which the causes of the modern peace can be reduced to any of its component parts. At best, each of them is a more or less adequate but still an imperfect proxy for a much broader development.

The various elements of the Modernization Peace account for the Long Peace phenomenon, occurring three consecutive times and for ever longer periods from

1815 onward. Indeed, the Long Peace phenomenon encompasses both the nine-teenth century's well-recognized pacificity and the post-1945 decline of war within a comprehensive explanatory framework. Similarly, the asymmetrical unfolding of the Modernization Peace explains why war has all but died out within the devel-oped parts of the world, has decreased dramatically in the developing parts, and mainly survives in the least developed parts. Aggregate measurements of global belligerency often fail to reflect the full extent of the steep drop in belligerency within the developed world, and, more importantly, leave unspecified the underly-ing factor responsible for the entire trend. The COW database, limited as it is to the period from 1816 onward, has similarly obscured the extent and cause of the change in comparison to earlier times. The twenty-first century will offer further testing ground for the Modernization Peace, particularly with respect to the con-tinued modernization of China and the development of presently poorly developed parts of the world, mostly in Africa and Asia.

Attempts to find cyclical patterns of war occurrence (all unsuccessful) and real-ist theorizing both assume an unchanging reality and fail to grasp how fundamental the modern transformation has been. The claim that everything stays basically the same even in the most modernized and most deeply pacified parts of today's world such as Western Europe or North America is patently untenable. In these parts, the security dilemma itself, that most elementary phenomenon of international rela-tions throughout history, has completely disappeared, an entirely unprecedented occurrence. Realists have never been able to explain why Holland and Belgium no longer have the slightest concern regarding the possibility of a German (or French) invasion, or what exactly defends Canada from conquest by the United States and accounts for its complacency in the face of such a prospect. In East Asia, the most developed countries, such as Japan, South Korea, and Taiwan, do not fear war among themselves or with any of the other developed countries, though they are deeply apprehensive of being attacked by less developed neighbours, such as China or North Korea.

This is not to say that regressions and relapses in the modernization processes and, as a result, also in the Modernization Peace cannot occur. Such relapses have happened at an earlier stage of modernization, during the world war period, and an unravelling of the process is not inconceivable even at a much more advanced stage. The point is different: the Modernization Peace represents as radical a change from earlier history as the modern transformation itself; and for it to fail requires a breakdown of modernization itself or serious setbacks in its major aspects as they are currently known.

On the other hand, while liberal theorists have been fundamentally correct in analyzing the transformative nature of the changes that have occurred since the beginning of the nineteenth century, there is always the peril of underplaying the process's considerable limitations and constraints. The international environment is still very challenging, largely conflictual, and often dangerous, where the use of force cannot be ruled out and is sometimes necessary. Realism retains credence and value as a warning against idealist illusions with respect to rivals that do not conform to the full spectrum of the Modernization Peace, above all, today, China and Russia. This, however, does not mean that great powers relations remain basically the same, as realists presage. Rather, what it means is that the validity of the Modernization Peace depends on the level and scope of modernization. What so conspicuously applies to North America, Western Europe, and Japan, the most advanced modernizers and the most liberal and democratic countries, may not apply or apply less effectively to China and Russia. Not only are these countries still far behind on the road to modernization; it also remains to be seen how sustainable their alternative models of modernization will be, with their resistance to democratization and liberalization – highly potent elements of the Modernization Peace.

Overall, there is room for guarded optimism. Modernization is an extremely powerful trend and will need extraordinary forces to derail. Nonetheless, ethnonational tensions are a deep source of potential armed conflict, especially in the less developed parts of the world and where political and ethnic borders do not cohere. Moreover, alternative routes to modernity are back on offer and competing in the world arena, with significant policy implications. In addition, both antimodernism and failed modernism are still evident, sometimes assuming fanatical and gruesome expressions. While the modern transformation decreases the likelihood of war between states, the potential democratization and individualization of mass death and disruption may become a real and disconcerting prospect.

Notes

1 John Gaddis, *The Long Peace: Inquiries into the History of the Cold War* (Oxford: Oxford UP, 1989).
2 Pitirim Sorokin, *Social and Cultural Dynamics, Vol. 3: Fluctuation of Social Relationships, War, and Revolution* (New York: The Bedminster Press, 1962); Quincy Wright, *A Study of War* (Chicago: U. of Chicago P., 1965); Jack Levy, *War in the Modern Great Power System, 1495–1975* (Louisville: UP. of Kentucky, 1983); for the COW bias and that of other studies that begin in 1816, p. 113; Evan Luard, *War in International Society* (London: Tauris, 1986); Kalevi Holsti, *Peace and War: Armed Conflicts and International Order, 1648–1989* (Cambridge: Cambridge UP, 1991).
3 Wright, *A Study of War*, p. 653; Luard, *War in International Society*, chap.2, esp. pp. 24–5, 35, 45, 53, and appendices 1–4.
4 Jack Levy and William Thompson, *The Arc of War: Origins, Escalation, and Transformation* (Chicago: U. of Chicago P., 2011), pp. 15, 144–5.
5 Levy, *War in the Modern Great Power System*.
6 Ibid., p. 110.
7 Ibid., pp. 138–44, quotation from p. 144; also Sorokin, *Social and Cultural Dynamics*, pp. 297, 341–2.
8 Victor Hanson, *A War Like No Other: How the Athenians and Spartans Fought the Peloponnesian War* (New York: Random House, 2005), pp. 10–11, 79–80, 82, 264, 296.
9 Peter Brunt, *Italian Manpower 225 B.C. – A.D. 14* (Oxford: Oxford UP, 1971).
10 Frederick Mote, 'Chinese Society under Mongol Rule,' in H. Franke and D. Twitchett (eds.), *The Cambridge History of China: Alien Regimes and Border States, 907–1368* (Cambridge: Cambridge UP, 1994), pp. 618–22.
11 Also Matthew White, *Atrocitology: Humanity's 100 Deadliest Achievements* (London: Canongate, 2011), summary on p. 529.
12 Peter Wilson, *The Thirty Years War: Europe's Tragedy* (Cambridge, MA: Harvard UP, 2009), pp. 786–8.

166   *Modernization Peace and current conflict*

13 Richard Bonney (ed.), *Economic Systems and State Finance* (Oxford: Oxford UP, 1995); idem., *The Rise of the Fiscal State in Europe, c.1200–1815* (Oxford: Oxford UP, 1999); Azar Gat, *War in Human Civilization* (Oxford: Oxford UP, 2006), pp. 371 and n. 92, 412, 472–6, 484–90.
14 John Mueller, *Retreat from Doomsday: The Obsolescence of Major War* (New York: Basic Books, 1989).
15 Carl Kaysen, 'Is War Obsolete? A Review Essay of Retreat from Doomsday: The Obsolescence of Major War,' *International Security*, 14:4 (1990), pp. 42–64.
16 Jean-Jacques Rousseau, 'Abstract and Judgement of Saint-Pierre's Project for Perpetual Peace' (1756), in S. Hoffmann and D. Fidler (eds.), *Rousseau on International Relations* (Oxford: Oxford UP, 1991), pp. 53–100; Thomas Paine, *Rights of Man, Common Sense, and Other Political Writings* (Oxford: Oxford UP, 1995); Immanuel Kant, 'Perpetual Peace: A Philosophical Sketch,' in H. Reiss (ed.), *Kant's Political Writings* (Cambridge: Cambridge UP, 1991), pp. 93–130.
17 Works on the subject are now legion. For a short list see: Dean Babst, 'A Force for Peace,' *Industrial Research*, 14 (Apr. 1972), pp. 55–8; Melvin Small and David Singer, 'The War-Proneness of Democratic Regimes, 1816–1965,' *Jerusalem Journal of International Relations*, 1:4 (1976), pp. 50–69; R.J. Rummel, 'Libertarianism and International Violence,' *Journal of Conflict Resolution*, 27 (1983), pp. 27–71; Michael Doyle, 'Kant, Liberal Legacies, and Foreign Affairs,' *Philosophy and Public Affairs*, 12 (1983), pp. 205–35, 323–53; William Domke, *War and the Changing Global System* (New Haven, CT: Yale UP, 1988); Zeev Maoz and Nasrin Abdolali, 'Regime Type and International Conflict, 1816–1976,' *Journal of Conflict Resolution*, 33 (1989), pp. 3–35; Zeev Maoz and Bruce Russett, 'Normative and Structural Causes of Democratic Peace, 1946–1986,' *American Political Science Review*, 87 (1993), pp. 624–38; Bruce Russett, *Grasping the Democratic Peace* (Princeton: Princeton UP, 1993); Michael Mousseau, 'Democracy and Compromise in Militarized Interstate Conflicts, 1816–1992,' *Journal of Conflict Resolution*, 42 (1998), pp. 210–30; Bruce Russett and John Oneal, *Triangulating Peace: Democracy, Interdependence and International Organizations* (New York: Norton, 2001).
18 See Azar Gat, 'The Democratic Peace Theory Reframed: The Impact of Modernity,' *World Politics*, 58, Oct. 2005, pp. 73–100, esp. pp. 80–3, for a broader discussion of Classical antiquity and the scholarly literature.
19 Livy, XXIII.ii–vii, XXIV.xiii.
20 Doyle, 'Kant, Liberal Legacies, and Foreign Affairs.'
21 Ibid.; Michael Doyle, *Ways of War and Peace: Realism, Liberalism, and Socialism* (New York: Norton, 1997); Russett and Oneal, *Triangulating Peace*.
22 Montesquieu, *The Spirit of the Laws* (Cambridge: Cambridge UP, 1989), Bk. 20, chap.2; Paine, *Rights of Man*, pp. 128–31, 227, 265–6; Kant, 'Perpetual Peace,' p. 114.
23 For some of the most important works, supportive, sceptical, or qualifying, see: Richard Rosecrance, *The Rise of the Trading State: Commerce and Conquest in the Modern World* (New York: Basic Books, 1986); Edward Mansfield, *Power, Trade, and War* (Princeton, NJ: Princeton UP, 1994); Erich Weede, 'Economic Policy and International Security: Rent-Seeking, Free Trade and Democratic Peace,' *European Journal of International Relations*, 1:4 (1995), pp. 519–37; Katherine Barbieri, 'Economic Interdependence: A Path to Peace or a Source of Interstate Conflict?,' *Journal of Peace Research*, 33 (1996), pp. 29–49; Katherine Barbieri and Gerald Schneider, 'Globalization and Peace: Assessing New Directions in the Study of Trade and Conflict,' *Journal of Peace Research*, 36 (1999), pp. 387–404; Solomon Polachek, 'Why Democracies Cooperate More and Fight Less: The Relationship between Trade and International Cooperation,' *Review of International Economics*, 5 (1997), pp. 295–309; Edward Mansfield and Brian Pollins (eds.), *Economic Interdependence and International Conflict* (Ann Arbor: U. of Michigan, 2003); G. Schneider, K. Barbieri, and N. Gleditsch (eds.), *Globalization and Armed Conflict* (Lanham, MD: Rowman & Littlefield, 2003); Russett and Oneal,

Triangulating Peace, pp. 125–55; Omar Keshk, Brian Pollins, and Rafael Reuveny, 'Trade Still Follows the Flag: The Primacy of Politics in a Simultaneous Model of Interdependence and Armed Conflict,' *Journal of Politics*, 66:4 (2004), pp. 1155–79; Jun Xiang, Xu Xiaohan, and George Keteku, 'Power: The Missing Link in the Trade-Conflict Relationship,' *Journal of Conflict Resolution*, 51:4 (2007), pp. 646–63; Zeev Maoz, 'The Effects of Strategic and Economic Interdependence on International Conflict across Levels of Analysis,' *American Journal of Political Science*, 53:1 (2009), pp. 223–40; Han Dorussen and Hugh Ward, 'Trade Networks and the Kantian Peace,' *Journal of Peace Research*, 47:1 (2010), pp. 29–42; Håvard Hegre, John Oneal, and Bruce Russett, 'Trade Does Promote Peace: New Simultaneous Estimation of the Reciprocal Effects of Trade and Conflict,' *Journal of Peace Research*, 47:6 (2010), pp. 763–74.

24 Erik Gartzke, 'The Capitalist Peace,' *American Journal of Political Science*, 51:1 (2007), pp. 166–91; also, Erik Gartzke and J.J. Hewitt, 'International Crises and the Capitalist Peace,' *International Interactions*, 36 (2010), pp. 115–45.

25 Bruce Russett, 'Capitalism or Democracy? Not So Fast,' *International Interactions*, 36:2 (2010), pp. 198–205;Allan Dafoe, 'Statistical Critiques of the Democratic Peace: Caveat Emptor,' *American Journal of Political Science*, 55:2 (2011), pp. 247–62; Allan Dafoe, John Oneal, and Bruce Russett, 'The Democratic Peace: Weighing the Evidence and Cautious Inference,' *International Studies Quarterly*, 57 (2013), pp. 201–14; also, Christopher Gelpi and Joseph Grieco, 'Democracy, Interdependence, and the Sources of the Liberal Peace,' *Journal of Peace Research*, 45:1 (2008), pp. 17–36.

26 B. Mitchell, *European Historical Statistics, 1750–1970* (London: Macmillan, 1975), pp. 526, 573; Kenneth Waltz, *Theory of International Politics* (Reading, MA: Addison, 1979), pp. 212–5.

27 Patrick McDonald, *The Invisible Hand of Peace: Capitalism, the War Machine, and International Relations Theory* (Cambridge: Cambridge UP, 2009); Patrick McDonald and Kevin Sweeney, 'The Achilles' Heel of Liberal IR Theory? Globalization and Conflict in the Pre – World War I Era,' *World Politics*, 59:3 (2007), pp. 370–403. A similar conclusion, albeit with somewhat different emphases, was reached by others earlier: Dale Copeland, 'Economic Interdependence and War: A Theory of Trade Expectations,'*International Security*, 20:4 (1996), pp. 5–41; Erich Weede, 'Globalization: Creative Destruction and the Prospect of a Capitalist Peace,' in Schneider, Barbieri, and Gleditsch, *Globalization and Armed Conflict*, pp. 311–23; Gat, *War in Human Civilization*, 2006, pp. 554–7, 585. See also Erik Gartzke and Lupu Yonatan, 'Trading on Preconceptions: Why World War I Was Not a Failure of Economic Interdependence,' *International Security*, 36:4 (2012), pp. 115–50.

28 Michael Mousseau, 'The Social Market Roots of Democratic Peace,' *International Security*, 33:4 (2009), pp. 52–86; idem., 'The Democratic Peace Unraveled: It's the Economy,' *International Studies Quarterly*, 57:1 (2012), pp. 186–97; Michael Mousseau et al., 'Capitalism and Peace: It's Keynes, Not Hayek,' in G. Schneider and N. Gleditsch (eds.), *Assessing the Capitalist Peace* (London: Routledge, 2013), pp. 80–109.

29 For the critics see Allan Dafoe and Bruce Russett, 'Does Capitalism Account for the Democratic Peace? The Evidence Still Says No,' in Schneider and Gleditsch, *Assessing the Capitalist Peace*, pp. 110–26. Håvard Hegre, 'Democracy and Armed Conflict,' *Journal of Peace Research*, 51 (2014), pp. 159–72, insists on the strong contributing effect of democracy.

30 Russett and Oneal, *Triangulating Peace*, suggest a distinct but mutually-reinforcing (rather than independent) effect to all three factors, which would seem to be the right answer for the industrial age. Similarly see Domke, *War and the Changing Global System*; Doyle, *Ways of War and Peace*, pp. 284, 286–7.

31 Kaysen, 'Is War Obsolete?,' is practically alone in pointing to the Industrial Revolution and the socio-economic changes it brought in its wake as the process behind the decline of war. Joseph Schumpeter, 'The Sociology of Imperialism,' in Joseph Schumpeter, *The*

Economics and Sociology of Capitalism (Princeton: Princeton UP, 1991), argued that capitalism, enhanced by the Industrial Revolution, was a force for peace.

32 Data on conflicts from: Uppsala University's Battle-Related Deaths Dataset; 'List of Ongoing Armed Conflicts,' *Wikipedia*; 'Modern Conflicts Table,' Political Economy Research Institute, University of Amherst; Joshua Goldstein, 'Wars in Progress,' July 2014, International Relations.com (all on the WWW). GDP per capita (nominal) from the United Nations, Statistics Division, 'National Accounts, Main Aggregates Database' (on the WWW).

33 For similar and other explanations see: Håvard Hegre et al., 'Predicting Armed Conflict, 2010–2050,' *International Studies Quarterly*, 57 (2013), pp. 250–70; James Fearon and David Laitin, 'Ethnicity, Insurgency, and Civil War,' *The American Political Science Review*, 97:1 (Feb., 2003), pp. 75–90.

34 Angus Maddison, *The World Economy: A Millennial Perspective* (Paris: OECD, 2001).

35 Auguste Comte, 'Plan of the Scientific Operations Necessary for Reorganizing Society' (1822), and 'Course de Philosophie Positive' (1832–1842), in G. Lenzer (ed.), *Auguste Comte and Positivism: The Essential Writings* (Chicago: U. of Chicago P., 1975), pp. 37, 293–7.

36 Helmuth von Moltke, *Essays, Speeches and Memoirs* (New York: HarperCollins, 1893), pp. i. 276–7.

37 Although Kaysen, 'Is War Obsolete?,' suggests that war has become intrinsically unprofitable in the industrial age, he notes Nazi Germany's exploitation of its European empire. This, as well as Japan's exploitation of its own empire, and a couple of other cases, are studied by Peter Liberman, *Does Conquest Pay? The Exploitation of Occupied Industrial Societies* (Princeton: Princeton UP, 1996), which answers in the positive. Stephen Brooks's criticism in his 'The Globalization of Production and the Changing Benefits of Conquest,' *Journal of Conflict Resolution*, 43:5 (1999), pp. 646–70; idem., *Producing Security: Multinational Corporations, Globalization, and the Changing Calculus of Conflict* (Princeton: Princeton UP, 2005), chap. 6, errs in choosing the Soviet empire as his case study. The empire may have been economically unbeneficial, but this was predominantly because, like the Soviet Union itself, it was communist, and therefore economically dysfunctional. Second, Brooks's claim that the wealth of countries which constitute part of a global production chain is difficult to retain under alien rule is not entirely persuasive. Consider the reluctant change of hands in Hong Kong, a major financial and trade hub of the information age.

38 As an afterthought to his random ideas thesis, Mueller has come to recognize the exponential rise in the levels of affluence since the beginning of the nineteenth century. However, he still espouses a one-directional relationship between ideas and material developments, arguing that changing ideas regarding prosperity, capitalism, and peace were the *cause* of rocketing affluence. He fails to grasp the deep interconnections between the Industrial Revolution, the growth of prosperity, and the changing attitudes. Mueller further argues that industrialization intensified the militaristic spirit during the nineteenth century, which was in fact the most pacific in history until then. He also argues that industrialization was responsible for the two world wars, forgetting that he blames at least the Second World War on Hitler alone. John Mueller, *The Remnants of War* (Ithaca: Cornell UP, 2007), pp. 33, 165–6; developed in his 'Capitalism, Peace, and the Historical Movement of Ideas,' in Schneider and Gleditsch, *Assessing the Capitalist Peace*, pp. 64–79.

39 Erik Gartzke and Alex Weisiger, 'Under Construction: Development, Democracy, and Difference as Determinants of Systemic Liberal Peace,' *International Studies Quarterly*, 58 (2014), pp. 130–45, suggest something along these lines. However, they make no sense in claiming that the developed countries themselves are hypocritical in that they continue to fight, whereas they practically ceased to fight among themselves.

40 Max Singer and Aaron Wildavsky, *The Real World Order: Zones of Peace, Zones of Turmoil* (Chatham, NJ: Chatham House, 1993); James Goldgeier and Michael McFaul,

'A Tale of Two Worlds: Core and Periphery in the Post-Cold War Era,' *International Organization*, 46 (1992), pp. 467–91; Paul Collier, *The Bottom Billion: Why the Poorest Countries are Failing and What Can Be Done about It* (Oxford: Oxford UP, 2007); B. Lacina and N.P. Gleditsch, 'Monitoring Trends in Global Combat: A New Dataset of Battle Deaths,' *European Journal of Population*, 21 (2005), pp. 145–66. Mueller, *The Remnants of War*, stresses the sharp difference between the developed and undeveloped world, but fails to draw the obvious conclusion regarding the key factor involved: economic development.

41 Joshua Goldstein, *Winning the War on War: The Decline of Armed Conflict Worldwide* (New York: Penguin, 2011). See also Andrew Mack, 'Global Political Violence: Explaining the Post-Cold War Decline,' in M. Fischer and V. Rittberger (eds.), *Strategies for Peace* (Leverkusen: Barbara Budrich, 2008), pp. 75–106.

42 Meredith Sarkees, Frank Wayman, and David Singer, 'Inter-State, Intra-State, and Extra-State Wars: A Comprehensive Look at Their Distribution over Time, 1816–1997,' *International Studies Quarterly*, 47:1 (2003), pp. 49–70, find little change in the overall occurrence of war during the period from 1816 onward. However, while noting the decrease of war in nineteenth century Europe and in today's developed world, they overlook the fact that most wars today, either external or internal, concentrate in the developing world. David Singer was more discriminate in his, 'Peace in the Global System: Displacement, Interregnum, or Transformation?,' in C. Kegley (ed.), *The Long Postwar Peace: Contending Explanations and Projections* (New York: HarperCollins, 1991), pp. 56–84, where he noted that wars between major powers as well as between 'advanced industrial nations' had decreased sharply.

43 Stuart Bremer, 'Dangerous Dyads: Conditions Affecting the Likelihood of Interstate War, 1816–1965,' *Journal of Conflict Resolution*, 36 (1992), pp. 309–41; Michael Mousseau, 'Market Prosperity, Democratic Consolidation, and Democratic Peace,' *Journal of Conflict Resolution*, 44 (2000), pp. 472–507; idem., 'The Nexus of Market Society, Liberal Preferences, and Democratic Peace,' *International Studies Quarterly*, 47 (2003), pp. 483–510; idem., 'Comparing New Theory with Prior Beliefs: Market Civilization and the Democratic Peace,' *Conflict Management and Peace Science*, 22 (2005), pp. 63–77; Håvard Hegre, 'Development and the Liberal Peace: What Does It Take to Be a Trading State?,' *Journal of Peace Research*, 37 (2000), pp. 5–30; Michael Mousseau, Håvard Hegre, and John Oneal, 'How the Wealth of Nations Conditions the Liberal Peace,' *European Journal of International Relations*, 9 (2003), pp. 277–314, the only study among those listed that extends to the period before World War II.

44 B.R. Mitchell, *International Historical Statistics, Europe, 1750–1988* (New York: Stockton, 1992), pp. 553–62; Maddison, *The World Economy*, pp. 126–7, 184; Simon Kuznets, *Modern Economic Growth* (New Haven, CT: Yale UP, 1966), pp. 306–7, 312–14.

45 John Stuart Mill, *Principles of Political Economy* (New York: Kelley, 1961), Book 3, chap. xvii, sect. 5, p. 582.

46 John Jackson, *The World Trading System* (Cambridge, MA: MIT P., 1997), p. 74.

47 Richard Rosecrance, *The Rise of the Virtual State: Wealth and Power in the Coming Century* (New York: Basic Books, 1999), p. 37; Robert Gilpin with Jean Gilpin, *The Challenge of Global Capitalism: The World Economy in the 21st Century* (Princeton: Princeton UP, 2000), pp. 20–3.

48 McKinsey Global Institute, 'Global Flows in a Digital Age: How Trade, Finance, People, and Data Connect the World Economy' (2014), pp. 4–5, 88 (on the WWW).

49 Smith, *The Wealth of Nations*, IV.ii.23.

50 Cf. Thomas Christensen, 'Fostering Stability or Creating a Monster? The Rise of China and U.S. Policy toward East Asia,' *International Security*, 31:1 (2006), pp. 81–126.

51 Data from the following sources (all on the WWW): US Census Bureau, 'US Trade: Top Trading Partners – Total Trade, Exports, Imports,' Dec. 2012; The European Commission, Directorate General for Trade, 'Top Trading Partners 2013'; World Trade

170   *Modernization Peace and current conflict*

Organization, 'Trade Profiles by Country'; *The CIA World Factbook 2013*, Country Comparison: Stock of Direct Foreign Investment.

52 McKinsey, 'Global Flows in a Digital Age,' pp. 4–5.

53 World Trade Organization, 'Trade Profiles by Country'; US Census Bureau, 'Top Trading Partners – December 2013'; European Commission, Directorate-General for Energy, 'Registration of Crude Oil Imports and Deliveries in the European Union, 1–12/2013'; European Commission, 'Quarterly Reports on European Gas Markets' (2014); Ralf Dickel et al., 'Reducing European Dependence on Russian Gas,' Oxford Institute for Energy Studies (2014); all on the WWW.

54 A recent exposition of this subject is Lilia Shevtsova, 'Forward to the Past in Russia,'*Journal of Democracy*, 26:2 (2015), pp. 22–36.

55 See Gat, 'The Democratic Peace Theory Reframed,' pp. 89–91, for some of the major features and effects of the rising affluence.

56 Ronald Inglehart and Christian Welzel, *Modernization, Cultural Change and Democracy: The Human Development Sequence* (New York: Cambridge UP, 2005).

57 Norbert Elias, *The Civilizing Process* (Oxford: Blackwell, 1994).

58 Gat, 'The Democratic Peace Theory Reframed,' pp. 91–2.

59 Rosecrance, *The Rise of the Virtual State*, pp. xii, 26; Robert Gilpin, *The Challenge of Global Capitalism* (Princeton, NJ: Princeton UP, 2000), p. 33.

60 The World Bank, 'Urban Population (% of Total)'; Wikipedia, 'Urbanization by Country' (both on the WWW).

61 The CIA, 'A Look at International Labor and Unemployment,' (on the WWW).

62 This helps to explain some of the (limited) finds in Mark Peceny, Caroline Beer, and Shannon Sanchez-Terry, 'Dictatorial Peace?,' *The American Political Science Review*, 96:1 (2002), pp. 15–26. Also, for the long-term decrease in the belligerency of non-democratic countries: Lars-Erik Cederman, 'Back to Kant,' *American Political Science Review*, 93:4 (2001), pp. 791–808.

63 For the correlation between the level of liberalism and peace, see Rummel, 'Libertarianism and International Violence'; R.J. Rummel, *Power Kills: Democracy as a Method of Nonviolence* (New Brunswick, NJ: Transaction, 1997), p. 5 and chap. 3. For historical gradualism, see Zeev Maoz, 'The Controversy over the Democratic Peace: Rearguard Action or Cracks in the Wall?,' *International Security*, 22:1 (1997), pp. 162–98; Russett and Oneal, *Triangulating Peace*, pp. 111–14.

64 Seymour Lipset, *Political Man* (New York: Anchor, 1963); Samuel Huntington, *The Third Wave: Democratization in the Late Twentieth Century* (Norman: U. of Oklahoma, 1991); Francis Fukuyama, *The End of History and the Last Man* (New York: Free Press, 1992); Larry Diamond, *Developing Democracy* (Baltimore, MD: Johns Hopkins UP, 1999); Amartya Sen, *Development and Freedom* (New York: Knopf, 1999); Mancur Olson, *Power and Prosperity: Outgrowing Communist and Capitalist Dictatorships* (New York: Basic Books, 2000); Adam Przeworski et al., *Democracy and Development* (Cambridge: Cambridge UP, 2000).

65 This is the Achilles heel of Bruce Bueno de Mesquita, James Morrow, Randolph Siverson, and Alastair Smith, 'An Institutional Explanation of the Democratic Peace,' *The American Political Science Review*, 93:4 (1999), pp. 791–807.

66 Rummel, *Power Kills*; Mathew Krain and Marrissa Myers, 'Democracy and Civil War: A Note on the Democratic Peace Proposition,' *International Interaction*, 23 (1997), pp. 109–18, which fails to distinguish between advanced and less advanced democracies; Tanja Ellingson, 'Colorful Community or Ethnic Witches-Brew? Multiethnicity and Domestic Conflict during and after the Cold War,' *Journal of Conflict Resolution*, 44 (2000), pp. 228–49; Errol Henderson and David Singer, 'Civil War in the Post-Colonial World, 1946–1992,' *Journal of Peace Research*, 37 (2000), pp. 275–99; Errol Henderson, *Democracy and War: The End of an Illusion* (Boulder, CO: Lynne Rienner, 2002),

chap. 5; Håvard Hegre, Tanja Ellingsen, Scott Gates, and Nils Petter Gleditsch, 'Toward a Democratic Civil Peace? Democracy, Political Change, and Civil War, 1816–1992,' *American Political Science Review*, 95:1 (2001), pp. 33–48; James Fearon and David Laitin, 'Ethnicity, Insurgency, and Civil War,' *American Political Science Review*, 97:1 (2003), pp. 75–90; Kristine Eck and Lisa Hultman, 'One-Sided Violence against Civilians in War,' *Journal of Peace Research*, 44:2 (2007), pp. 233–46; Christian Davenport, *State Repression and the Domestic Democratic Peace* (Cambridge: Cambridge UP, 2007); Nils Petter Gleditsch, Håvard Hegre, and Håvard Strand, 'Democracy and Civil War,' in M. Manus (ed.), *Handbook of War Studies III* (Ann Arbor: U. of Michigan, 2009), pp. 155–92.

67 Cf. Francis Fukuyama, *State Building: Governance and World Order in the 21st Century* (Ithaca, NY: Cornell UP, 2004), pp. 38–9, 92–3.

68 For the Soviets, see Anthony Beevor, *The Fall of Berlin 1945* (New York: Penguin, 2003), p. 410. For the Americans and Japanese in World War II, see Joshua Goldstein, *War and Gender: How Gender Shapes the War System and Vice Versa* (New York: Cambridge UP, 2001), pp. 337, 346.

69 Herbert Moller, 'Youth as a Force in the Modern World,' *Comparative Studies in Society and History*, 10 (1967–1968), pp. 237–60; Christian Mesquida and Neil Wiener, 'Human Collective Aggression: A Behavioral Ecology Perspective,' *Ethology and Sociobiology*, 17 (1996), pp. 247–62; Henrik Urdal, 'A Clash of Generations? Youth Bulges and Political Violence,' *International Studies Quarterly*, 50 (2006), pp. 607–29.

70 Mitchell, *European Historical Statistics*, sect. B2, esp. pp. 37, 52; United Nations, *World Population Prospects: The 2000 Revision* (New York: UN, 2001).

71 United Nations, *World Population Prospects: The 2000 Revision, the 2012 Revision* (New York: UN, 2013), Table S.8.

72 The World Bank, 'Fertility Rates, Total (Births per Woman),' (on the WWW); A. Korotayev et al., 'A Trap at the Escape from the Trap? Demographic-Structural Factors of Political Instability in Modern Africa and West Asia,'*Cliodynamics*,2:2 (2011), pp. 1–28.

73 Lisa Brandes, 'Public Opinion, International Security and Gender: The United States and Great Britain since 1945,' unpublished doctoral dissertation, Yale University, 1994.

74 Bruce Russet, 'The Democratic Peace: And Yet It Moves,' in M. Brown, S. Lynn-Jones, and S. Miller (eds.), *Debating the Democratic Peace* (Cambridge, MA: MIT P., 1996), p. 340; Michael Doyle, 'Michael Doyle on the Democratic Peace: Again,' *op. cit.*, p. 372.

75 The CIA, 'The World Factbook: Sex Ratio,' (on the WWW).

76 McKinsey, 'Global Flows in a Digital Age'; Fareed Zakaria, *The Post American World* (New York: Norton, 2008), p. 83.

77 Kalevi Holsti, *Peace and War: Armed Conflict and International Order, 1648–1989* (Cambridge: Cambridge UP, 1991), pp. 139–45.

78 Azar Gat, with Alexander Yakobson, *Nations: The Long History and Deep Roots of Political Ethnicity and Nationalism* (Cambridge: Cambridge UP, 2013), chap. 6; Benjamin Miller, *States, Nations and Great Powers: The Sources of Regional War and Peace* (Cambridge: Cambridge UP, 2007).

79 Cf. Dale Copeland, 'Economic Interdependence and War: A Theory of Trade Expectations,' *International Security*, 20:4 (1996), pp. 5–41.

80 Azar Gat, 'The Return of Authoritarian Great Powers,' *Foreign Affairs*, 86:4 (2007), pp. 59–69; idem., 'Which Way Is History Marching: Debating the Authoritarian Revival,' *Foreign Affairs*, 88:4 (2009), pp. 150–5; idem., *Victorious and Vulnerable: Why Democracy Won in the 20th Century and How It Is Still Imperiled* (Stanford: Hoover/Rowman & Littlefield, 2010); Larry Diamond, 'Facing Up to the Democratic Recession,'*Journal of Democracy*,26:1 (2015), pp. 141–55, and the other contribution to that issue.

81 Sketching possible scenarios and stressing that it is impossible to know which of them will eventually materialize are: Aaron Friedberg, 'The Future of U.S.-China Relations: Is Conflict Inevitable?,' *International Security*, 30:2 (2005), pp. 7–45; idem., *The Contest for Supremacy: China, America, and the Struggle for Supremacy in Asia* (New York: Norton, 2011); Legro Jeffrey, 'What China Will Want: The Future Intentions of a Rising Power,' *Perspective on Politics*, 5:3 (2007), pp. 515–34; Randall Schweller and Xiaoyo Pu, 'After Unipolarity: China's Visions of International Order in an Era of U.S. Decline,' *International Security*, 36:1 (2011), pp. 41–72; Henry Kissinger, *On China* (New York: Penguin, 2011); Edward Steinfeld, *Playing Our Game: Why China's Economic Rise Doesn't Threaten the West* (Oxford: Oxford UP, 2010), is generally optimistic. Yong Deng, *China's Struggle for Status: The Realignment of International Relations* (Cambridge: Cambridge UP, 2008); David Shambaugh, *China Goes Global: The Partial Power* (Oxford: Oxford UP, 2013), are useful surveys.
82 John Mueller, *Overblown: How Politicians and the Terrorism Industry Inflate National Security Threats, and Why We Believe Them* (New York: Free Press, 2006).
83 Ibid., p. 20.

Index